E-manufacture

PEARSON
Education

We work with leading authors to develop the
strongest educational materials in engineering,
bringing cutting-edge thinking and best learning
practice to a global market.

Under a range of well-known imprints, including
Prentice Hall, we craft high quality print and electronic
publications which help readers to understand and
apply their content, whether studying or at work.

To find out more about the complete range of our
publishing, please visit us on the World Wide Web at:
www.pearsoneduc.com

E-manufacture

Application of Advanced Technology to Manufacturing Processes

Roger Timings
Steve Wilkinson

PEARSON

Prentice
Hall

Harlow, England • London • New York • Boston • San Francisco • Toronto • Sydney • Singapore • Hong Kong
Tokyo • Seoul • Taipei • New Delhi • Cape Town • Madrid • Mexico City • Amsterdam • Munich • Paris • Milan

Pearson Education Limited
Edinburgh Gate
Harlow
Essex CM20 2JE
England

and Associated Companies throughout the world

Visit us on the World Wide Web at:
www.pearsoneduc.com

First published 2003

© Roger Timings and Steve Wilkinson 2003

ISBN 0 582 43230 8

British Library Cataloguing-in-Publication Data
A catalogue record for this book is available from the British Library

Library of Congress Cataloging-in-Publication Data
Timings, R. L. (Roger Leslie)
    E-manufacture / Roger Timings, Steve Wilkinson.
        p. cm.
    Includes index.
    ISBN 0-582-43230-8 (pbk.)
        1. Production engineering. 2. Manufacturing processes. 3. Computer integrated
    manufacturing systems. I. Wilkinson, Steve (Steve P.) II. Title.

    TS176.T48 2003
    670'.285—dc21   2003040559

10 9 8 7 6 5 4 3 2 1
08 07 06 05 04 03

Typeset in 10/12 Times and Frutiger by 35
Produced by Pearson Education Malaysia Sdn Bhd,
Printed in Malaysia

The publisher's policy is to use paper manufactured from sustainable forests.

Contents

Chapter 7 The future of manufacturing

Preface

E-manufacturing is the application of Internet and intranet technologies to the design, marketing and manufacturing of products, together with the management and control of manufacturing industries.

E-manufacturing is essentially a free standing text for the new design and information technology (IT) based degree courses that students find increasingly attractive. It is written to provide an understanding of the application of IT to the practice and management of modern manufacture, and it is designed to satisfy the requirements of students studying for the new technology based foundation degree courses that embrace IT as part of their core.

In view of the ever-broadening spread of modern technology this text is based on those elements that form essential transferable skills and are appropriate across a wide range of courses. These aims are achieved by examining and discussing:

- the Web-based technologies as applied to the organisation and management of manufacturing systems;
- the automation and control of manufacturing systems;
- the new tools and techniques for the rapid design, development and production of prototype products (e.g. virtual products and rapid prototyping);
- machines and techniques to reduce lead times and provide for agile manufacturing (e.g. metal forming).

Each chapter is prefaced by a summary of topic areas to be covered and finishes with a selection of assignments. It is assumed, in some of the assignments, that students using this book will have access to relevant information that is available on the World Wide Web.

Those students who wish to apply the techniques expounded in this text to mechanical engineering manufacture are recommended to consult *Manufacturing Technology*, Volume 1 (Timings) and *Manufacturing Technology*, Volume 2 (Timings and Wilkinson), which cover conventional and computer aided manufacturing processes. Also recommended are *Engineering Materials*, Volumes 1 and 2 (Timings), because a thorough understanding of the available materials and their properties is essential to the design process. The appendix to this book, *E-manufacture*, lists the contents of these other books for your reference.

R. L. Timings
S. P. Wilkinson
2003

Acknowledgements

The authors and publishers are grateful to the following organisations for permission to reproduce their copyright material:

Sony United Kingdom Ltd for Figs 1.1, 1.14, 1.15, 1.17 and 1.25, 'Sony' and 'Walkman' are registered trademarks of Sony Corporation, Japan; AutoDesk Inc. for Figs 1.2, 1.3, 1.4, 1.5, 1.6, 1.7, 1.8, 1.9, 1.10, 1.11, 1.12 and 1.13; Cycore Inc. for Figs 1.18, 1.19, 1.20, 1.21, 1.22, 1.23, 1.25 and 1.26; VARI-FORM Inc. for Figs 1.26, 4.22, 4.23, 4.24, 4.25 and 4.26; Pearson Education for Figs 2.9, 2.10, 3.1–3.35, 4.10–4.12, 4.14, 5.15–5.21, 6.1–6.17, 7.1–7.3 and Tables 3.1, 3.2 and 4.1; DeskProto Inc. for Figs 2.11 and 2.12; Jones and Shipman International for Figs 4.1, 4.2, 4.3, 4.4, 4.5, 4.6 and 4.7; Centreless Tooling Co. Ltd for Figs 4.8 and 4.13; Amada United Kingdom Ltd for Figs 4.15, 4.16, 4.17, 4.18, 4.19, 4.20 and 4.21, © Amada United Kingdom International Limited; Eagle Precision Technologies Inc. for Figs 4.27, 4.28, 4.29, 4.30, 4.31, 4.32, 4.33, 4.34, 4.35, 4.36, 4.37 and 4.38; SCM Group (UK) Ltd for Figs 4.39 and 4.40; Colter Group Ltd for Figs 5.5, 5.6 and Table 5.1; Professor Nabil Gindy (Nottingham University)/Giddings and Lewis Inc. for Figs 7.7 and 7.8.

Part A
E-design

1 The influence of information technology on design

When you have read this chapter you should be able to:

- appreciate marketing and design philosophies;
- design a virtual product;
- model using three-dimensional geometry;
- understand the creation of interactions for virtual products;
- place a virtual product on a Web site;
- engineer real product designs successfully;
- appreciate the benefits of design and concurrent engineering.

1.1 Introduction

The basic principles of design specification and design for manufacture were covered in *Manufacturing Technology*, Volume 1 and *Manufacturing Technology*, Volume 2, by the authors of this book. Many different processes were described, and their selection criteria using *design for manufacture and assembly* (DFMA) was analysed. The management of manufacture was also covered from detailed process planning to safety standards. Also covered were items on advanced manufacturing technology, which was mainly driven by the advent of computer control from computer numerical control (CNC) and robotics to flexible manufacturing systems. This book builds on that information to describe how manufacturing, engineering and business is affected and controlled via the World Wide Web, the Internet and other important advances in information technology.

1.2 Marketing and design philosophy

The market place is characterised by the need for increasingly frequent product variants each having a shorter design-life cycle than the previous one. Therefore lead-time demands are shorter than ever. Before a company commits itself to the manufacture of a new product, a thorough market survey and product testing is required. This is standard marketing practice used in order to arrive at a final design. Once the product has been market tested a number of times, the design can be refined. This involves redesigning the product in light of comments made by the design team and potential customers. This process is called the *design iteration* where market analyses and refinements are repeated until the optimal solution is achieved.

1.3 Virtual products

In a trading environment it is imperative that products are advertised and marketed globally. To transport samples of full-size machines to prospective clients is unrealistic in terms of the physical effort and costs involved. Before the advent of information and communications technology (ICT), scale models were constructed for the use of sales personnel but this also was costly and time consuming. In an age when business depends upon instant communications and information, the solution is to create simulated designs of virtual products that can be transmitted globally via the Web. Such simulations employ the following features:

1. True-scale representations of the actual product modelled in three-dimensional solid geometry and in true colour.
2. Solid models that are photorealistic. That is, they have properties such as translucency, opacity and radiosity.
3. Bit maps that can be captured from the real product and mapped onto the appropriate faces of the virtual product. For example company logos, safety information and material textures using 'bump maps'.
4. They can have multimedia features such as:
   (a) audio – sound effects simulating the noise the mechanism makes, voice over sales promotion and background music;
   (b) animation – to bring the model to life so that it can be seen in operation;
   (c) special effects – such as smoke, flames, water, and other particle and fluid systems;
   (d) gravitation effects – spring mass damper systems;
   (e) interactivity – allows a potential customer to have hands-on experience of the functionality of the product.

The main method of projecting products across the globe is the *Internet*. Let us now examine how virtual, interactive products can be developed.

1.3.1 *Three-dimensional geometry*

There are many commercial software packages on the market that can be used to develop three-dimensional (3D) designs of products. For example ProEngineer, SolidEdge, AutoCad and 3D Studio Max.

There is not room to examine all of these examples of 3D design software packages. The examples in this chapter and throughout this book are based on 3D Studio Max. This is a universal package that can be used to design and develop products ranging from games to virtual walk-through architecture.

As an example of a virtual product, let us now consider the modelling of a typical cassette player, such as the Sony Walkman, using 3D Studio Max. To do this, the following techniques were used:

- *Editable Meshes* for the case.
- *Extrusions* for the buttons and lid.
- *Mesh Smooth* to create realistic surface finish.
- *Exact Pivot* points with revolute joints for hinges.
- *Sliding Joints* for mechanisms.

1.3.2 Photorealism

To make a product truly virtual, all colours, materials and surface finish effects can be mapped directly onto the appropriate face of the object. For example architects can map images of windows and doors directly onto each face of a simple cube to give a feeling of effect and detail quickly and easily. The same is true for equipment where logos, buttons and lights can be mapped onto appropriate faces without the need to develop the geometry for these objects. For example let us consider a Sony Walkman cassette player, as shown in Fig. 1.1.

1.3.3 Web programming software

A typical Web development software is called Cult 3D. It is used in the case study at the end of this chapter. In the mean time let us consider how Cult 3D is used. This package can read geometry generated within 3D Studio Max, providing that an appropriate, additional piece of software (referred to as a 'plug-in') is installed in the Studio Max directory. Within Cult 3D software, interactions can be developed in much the same way as 'hot spots' within Web pages. When you click the mouse button on an icon on a Web page this will generate an interaction. This is because the icon, in this example, is a 'hot spot'. Within Cult 3D, the imported geometry can be specified as a 'hot spot' or icon. The icons are 3D objects.

The Cult 3D system has its own visual programming system that uses icons to represent *actions* and *events* that are sequenced together by the use of *visual connections*. For example an *object* (the *button*) that, when clicked by the mouse (the *event*), will cause an *action* to occur. If the virtual button is virtually pressed, then the lid will appear to open (*revolute movement*). Also, all view points, i.e. rotational (revolute) and zoom functions, can be programmed into the object using Cult 3D.

1.3.4 Exporting to the Web and viewing the product

When the virtual product has been programmed to be fully interactive, a Web page can then be exported to the Internet. This page will appear to have all the attributes that were generated within Cult 3D and can be viewed using any Web page viewing tools. This Web page can now be up loaded on to the Web site using an appropriate Web site server by means of *file transfer protocol* (FTP). A Web address is given to identify its location.

Fig. 1.1 *A typical cassette player*

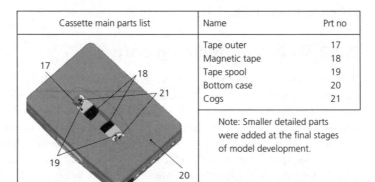

Cassette main parts List	Name	Prt no
	Top case	1
	Various text objects	2
	Battery cover	3
	Middle section (sandwich)	4
	Fast forward button	5
	Rewind button	6
	Hold button	7
	Stop button	8
	Play button	9
	Dolby button	10
	Bass button	11
	Volume control	12
	Volume bar	13
	Head phone socket	14
	Open button	15
	Logo	16

Cassette main parts list	Name	Prt no
	Tape outer	17
	Magnetic tape	18
	Tape spool	19
	Bottom case	20
	Cogs	21

Note: Smaller detailed parts were added at the final stages of model development.

The advantages of virtual products can be summarised as follows:

- Global availability.
- Collaborative design within a global company.
- The product can be viewed from any angle.
- The ability to zoom into any feature.
- Real material effects can be examined.
- A 'feel' for the product, i.e. operate its mechanisms, and operate the device to appreciate its audio and/or visual outputs, can be obtained.
- There can be interaction with the design, i.e. the colour or a detail of the product, such as the style of the wheels on a virtual car, can be changed.
- An object can be opened and entered to examine more detail. For example opening the door of a virtual car to examine the interior layout and trim, or opening the bonnet of a virtual car to examine the engine compartment layout.

1.4 The development of a virtual product

The Sony Walkman cassette player shown in Fig. 1.1 can be created as a virtual product by using 3D Studio Max and Cult 3D. This cassette player was chosen because of its potential to produce an interactive virtual product that could play music and allow the user of the Web page to interact and observe operational features.

The planned features for the *virtual* cassette player include the following:

1. An interactive play and stop button that can be activated using the mouse. By activating the play button a sound file can be played and the user can also view the tape spools rotating through the back window of the cassette player.
2. An interactive eject button can be used to open the cassette door by pressing the 'open' button on the front of the cassette player.
3. Interactive switches and volume control (linked to the sound quality, if possible).

1.4.1 *Using 3D Studio Max*

Before the cassette player can be made into an interactive product a great deal of time must be spent on planning and modelling the product. The chosen software package used to accomplish the modelling task is 3D Studio Max. This package is used for a wide variety of commercial and artistic applications and is much more than just a 3D modelling package. It offers the user the ability to create complex scenes that include models, animation, camera views, lighting and material effects. In short, it gives the user the ability to create an entire virtual world or environment.

1.4.2 *Modelling the cassette player – the initial commands*

The first task in producing a model of the cassette player is to determine and identify each of the components that make up the Walkman. By producing a parts list the cassette player can be broken down into smaller manageable parts that can be modelled up separately and merged in one complete scene. This was shown in Fig. 1.1.

Extrude command

To produce the finished model of the cassette player, many different techniques are utilised. One technique used throughout this example of the virtual construction of the cassette player is the *extrude* command. This command enables simple two-dimensional (2D) wire profiles to be extruded in the Z-axis to create an object with thickness, as shown in Fig. 1.2. Most parts of the cassette player are initially created using this technique. More specialised techniques can then be used to further define each part.

Bevel command

Once an object has been extruded, it can be converted into an *editable mesh*. This allows the user to manipulate the object to a much greater degree. For example, at a sub-object

Fig. 1.2 *Extrude command*

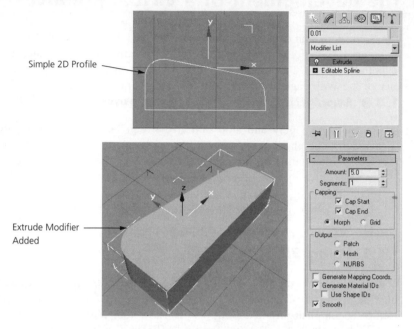

Simple 2D Profile

Extrude Modifier
Added

Fig. 1.3 *Adding the bevel*

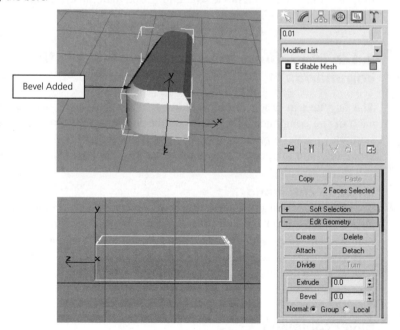

Bevel Added

level, individual vertices, edges and polygons can be manipulated. The *bevel* command is the technique used in the creation of the virtual buttons of the cassette player. This allows an angled chamfer to be constructed around the edge (sub-object) of the button. By selecting the top face of the object and selecting the bevel command, the top face can be shaped, as shown in Fig. 1.3.

1.4.3 *Modelling the cassette player – Boolean operations*

One very important technique used throughout the modelling of the virtual cassette player is the **Boolean operation** command. Boolean algebra was invented by the British mathematician George Boolean for the purpose of combining and manipulating sets of mathematical symbols. In this application, Boolean operations are applied to sets of surface meshes or solids, and enable the user to add and subtract mass by combining separate objects and selecting a suitable Boolean operation. The Boolean command consists of three main options. These are *Union*, *Subtraction* and *Intersection*. The Boolean commands enable the user to:

- take one solid away from another to form a single solid (*Subtraction*);
- combine separate solids together to form one complete solid (*Union*);
- produce a single solid made up from the intersection of two separate solids (*Intersection*).

Let us now look at subtraction and union operations more closely in the development of some components used in the cassette model.

Subtraction
In this example the objects during a Boolean command are referred to as *operand A* and *operand B*. To illustrate how the subtraction Boolean operation is used to create the back case, see Figs 1.4–1.7.

The sequence of operations used to subtract **operand A** from **operand B** in order to form a hole in the cassette lid is shown in Fig. 1.4. The outcome of this subtraction is shown in Fig. 1.5.

The same technique can be used to produce the hole in the back of the case as shown in Figs 1.6. and 1.7.

The subtraction operand is also used extensively to produce the button depression pockets that are on the side of the virtual cassette player. These include the Dolby button, bass button and volume control housing. The result of the pocket subtractions is shown in Fig. 1.8.

Union
The union operation is also used throughout the construction of the cassette player. This option enables two separate solids to be joined together to produce one complete solid, hence the term *union*. Figure 1.9 shows how the union technique is used to create the 'inner body' part of the cassette player using several Boolean union operations.

Fig. 1.4 *The Boolean operation for the recess*

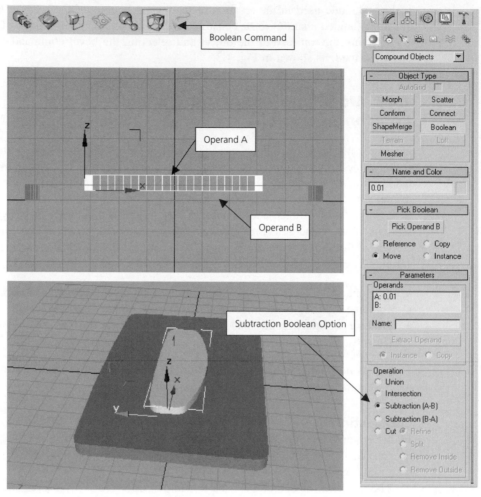

1.4.4 *Manipulation of editable meshes*

Any solid object can be converted to an *editable mesh*. Once it is converted, the user can manipulate the object at the sub-object level using a variety of control points, as shown in Fig. 1.10.

The sub-object menu shows the vertex command highlighted. The user can now select the vertices that need to be manipulated, so that they can be transformed in any axis to alter the original shape. This operation is shown in Fig. 1.11 and the effect of transforming the vertices in the Z-axis is shown in Fig. 1.12.

The domed shape on the top case of the cassette player model is constructed in this manner, although the transformation in the Z-axis is much less extreme. Selecting a pair

Fig. 1.5 *Outcome of the Boolean operation for the recess*

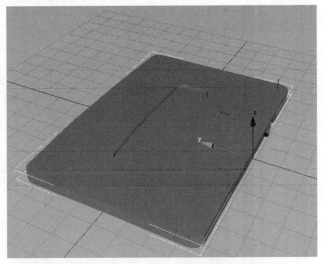

Fig. 1.6 *The Boolean operation for the cassette player*

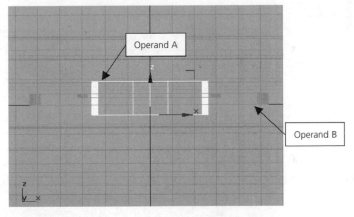

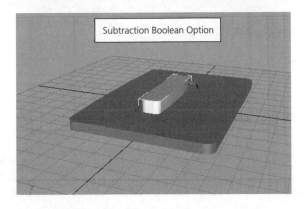

Fig. 1.7 *Outcome of the Boolean operation*

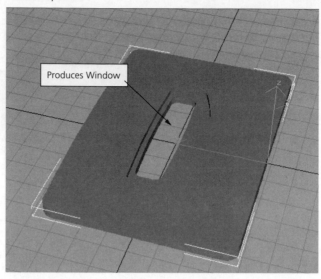

Produces Window

Fig. 1.8 *The side of the cassette player showing the depressions for the cassette controls*

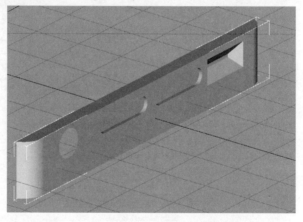

of vertices and transforming them in the Z-axis to form two small peaks creates the 'grip' on the button, as shown in Fig. 1.13.

1.4.5 *Text command*

This very useful technique, used to add detail to the modelling of the cassette player, is enabled through the use of the text command tool. Text is keyed into an input field so that it can be displayed as an *object* in the scene. Different fonts can be chosen and then manipulated accordingly, that is, size, scale and orientation, as shown in Fig. 1.14.

Fig. 1.9 *Union of inner body parts*

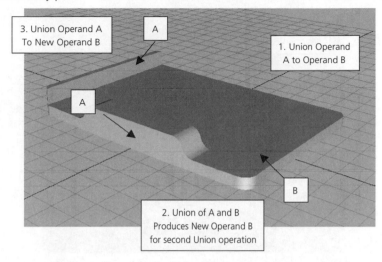

3. Union Operand A To New Operand B

A

1. Union Operand A to Operand B

A

B

2. Union of A and B Produces New Operand B for second Union operation

Fig. 1.10 *Editable meshes*

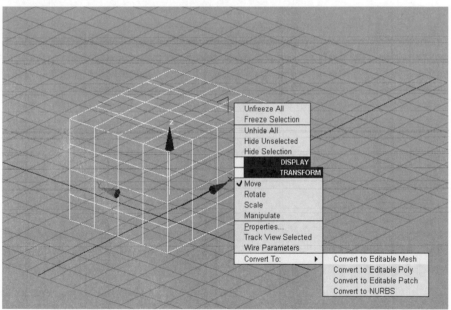

To create 3D lettering on the side of the modelled cassette player an appropriate logo or other information is keyed into the text input field to create a 2D *spline*. The *extrude modifier* is then added to give the letters a solid appearance that casts shadow on the finished *rendered* scene. Text can be manipulated in many ways, for example by the use of a non-uniform scaling factor, as shown in Fig. 1.15.

Fig. 1.11 *Selection of vertices*

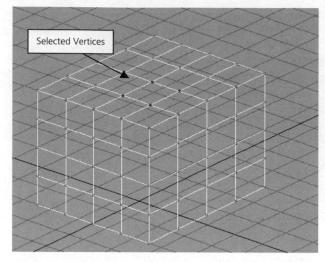

Fig. 1.12 *Transformation of vertices*

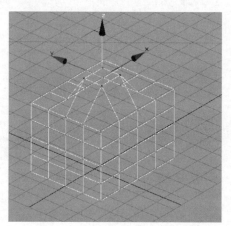

Fig. 1.13 *Sliding button 'grip'*

Fig. 1.14 *Text spline*

Fig. 1.15 *Text extrusion*

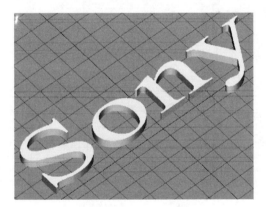

1.4.6 *Materials and mapping*

Once all of the separate parts of the cassette player have been modelled, the materials and maps can be applied to create a more 'photorealistic' object. Materials control how objects reflect and transmit light within a scene. They give the objects their final finish, that is, translucency, opacity and radiocity.

The icon shown in Fig. 1.16(a) brings up the material editor. Within the material editor is a menu called *Blinn Basic Parameters*. This enables the colour and reflective qualities of the material to be determined. Below are two examples of how to use the *specular highlight option*. Figure 1.16(b) shows an example of a glossy surface with a

Fig. 1.16 *Materials and mapping: (a) editor; (b) glossy; (c) matt*

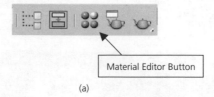

Material Editor Button

(a)

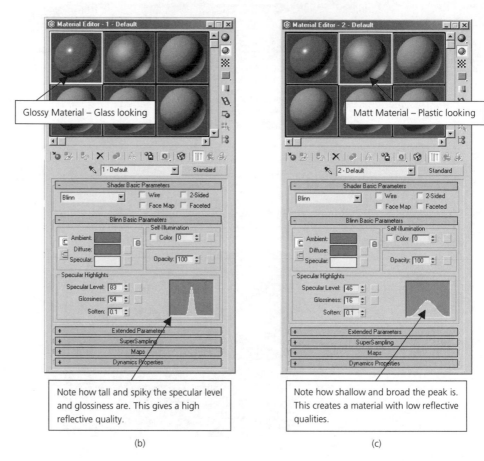

Glossy Material – Glass looking

Matt Material – Plastic looking

Note how tall and spiky the specular level and glossiness are. This gives a high reflective quality.

Note how shallow and broad the peak is. This creates a material with low reflective qualities.

(b)   (c)

tall and spiky specular level that has high reflective qualities, while Fig. 1.16(c) shows a matt surface with a short and broad specular level that has low reflective qualities. Since the modelled case of the cassette player is designed to mimic a plastic appearance, the material attributes shown in Fig. 1.16(c) were chosen.

1.4.7 *Mapping detail*

Mapping and materials go hand in hand. For the modelled cassette player it is necessary to map the name of the cassette player along with several serial number logos. To

Fig. 1.17 *Using bitmaps: (a) bitmap; (b) image*

(a)

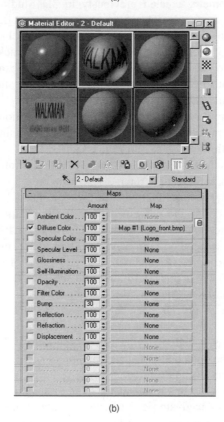

(b)

do this we first create a flat *bitmap* of the detail we want to use. Figure 1.17(a) shows the bitmap used. The background colour was set to be the same as the ambient colour in the material editor. By loading the bitmap it can be dragged and dropped onto a material. The image should then be seen as wrapped around the material sphere, as shown in Fig. 1.17(b).

The next stages are to:

- select the object to which the map is to be applied, that is the cassette player body;
- wrap the 2D map onto the 3D cassette player body, for which we have to specify a UVW map modifier (UVW defines the surfaces parallel to the- X-, Y- and Z axes). This tells the computer how to align the map onto the modelled surface so that it is not scrambled in appearance.

1.4.8 *Making the model interactive using Cult 3D software*

The next stage in the creation of the virtual product is to make the cassette player interactive. This stage of development is achieved by using the second software package called Cult 3D.

The geometry created previously in 3D Studio Max can now be exported as a Cult 3D designer file. This file can then be imported into Cult 3D as a new project. Cult 3D allows the user to interact with the virtual cassette player itself, together with its various buttons and objects. The Cult 3D package holds all of the information belonging to the 3D Studio Max geometry, for example the correct object hierarchy from 'parent to child'. If the hierarchy is incorrect, the movement and translation of the object in Cult 3D can be compromised.

Let us now consider programming the interactions to achieve this viewing interface (rotation, translation and zoom) and mechanism animation (opening the cassette player door – rotation), as shown in Fig. 1.18.

To open the cassette door we have to drag and drop the two objects from the geometry with which we want to interact (the open button and back case) as shown in Fig. 1.19.

Figure 1.19 shows that when we want to use a left mouse-click to activate the open button, we define the movement with the translation [command **1**] (to note the commands and icons in square brackets, refer to Fig 1.19). We link the 'open button' object icon to the translation command and then to the mouse event [icon **2**]. It is possible to move two or more objects by the activation of one mouse-button event. From the diagram shown in Fig. 1.19, we can see that the bottom case is also linked via a rotation [command **3**].

In order to prevent each left-hand mouse activation moving both objects in only one direction, an equal and opposite movement is required to close the case [icon **4**]. The problem is that by trying to open and close the cassette door simultaneously by activating the open button with the mouse we have a conflict of commands, which results in a door that does not move.

To overcome this problem we have to sequence the actions. By applying activate [icon **5**] and deactivate [icon **6**] commands and linking them correctly, we can tell the program to ignore one action until the first action is completed. This will let the 'open button' move and 'back case' open and close for every left-hand mouse activation.

To prevent the activation of the open button occurring through any left-hand click on any part of the cassette player we must close the loop by dragging the 'open button' link back to the mouse-activation icon [link **7**].

This same technique can be used for the movement of the bass button, the Dolby button, the hold button and the volume control wheel, as shown in Figs 1.20–1.23 inclusive.

1.4.9 *Interactive play – sounds*

The main interactive element for the cassette player is the ability for the user to press the play button and listen to a sample sound track. The following program, as shown in Fig. 1.24(a), allows this feature to be realised.

Fig. 1.18 *Viewing interaction using camera view, arcball and cassette object*

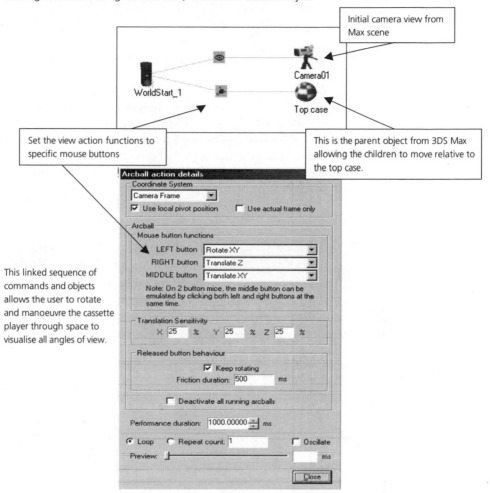

Initial camera view from Max scene

Camera01

Top case

WorldStart_1

Set the view action functions to specific mouse buttons

This is the parent object from 3DS Max allowing the children to move relative to the top case.

This linked sequence of commands and objects allows the user to rotate and manoeuvre the cassette player through space to visualise all angles of view.

Fig. 1.19 *Opening and closing the cassette player (interaction)*

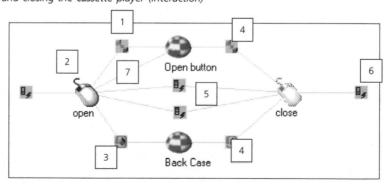

Fig. 1.20 *Operation of the bass button*

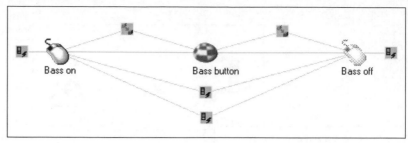

Fig. 1.21 *Operation of the Dolby button*

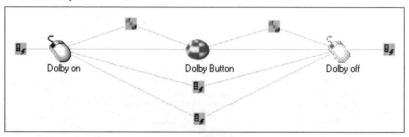

Fig. 1.22 *Operation of the hold button*

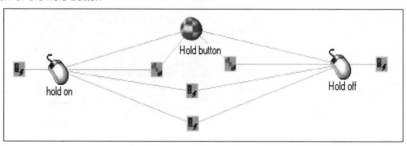

Fig. 1.23 *Operation of the volume control*

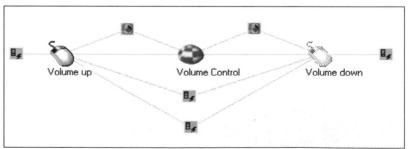

Fig. 1.24 *Interaction of the audio and the spool rotation: (a) operation of the play command; (b) duration of rotation of tape spool*

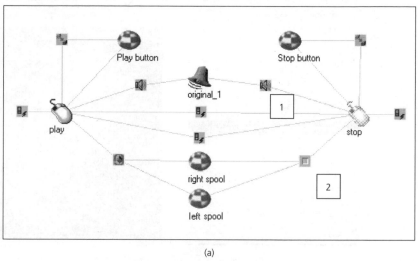

(a)

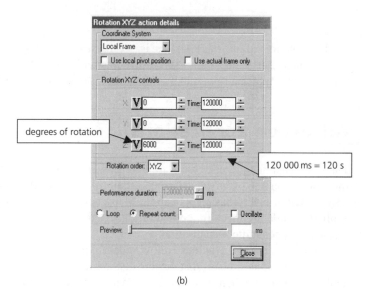

(b)

This program enables audio to be played when only the play button is pressed and also activates the rotation of the cassette tape spools. When only the stop button is activated the sound track will stop together with the cassette spool rotation. As previously, it is possible to link several different events to one left-hand, mouse-button activation. The sound track and rotation are linked to the play button and are stopped by the stop button. The music end command [icon **1**] and stop motion [icon **2**] allow both events to end with the stop button. The duration of the sound track is 120 seconds long, as shown in Fig. 1.24(b).

Fig. 1.25 *The finished Web page*

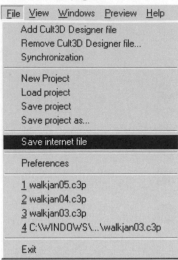

The rotation of the tape spools is set to stop precisely at the end of the track. The finished Cult project can then be saved as an Internet file for viewing, as shown in Fig. 1.25. All information including the sound sample should be in this file.

1.5 Advanced applications

Three-dimensional graphics tools are very powerful software packages that can be used for a wide variety of advanced commercial and artistic applications. So far we have only looked at a very basic application. Many applications to which 3D graphics can be applied include architecture, computer games, film production, Web design, forensics, medical and scientific visualisation.

Three-dimensional graphics is now much more than just solid modelling. The ability to animate scenes means that objects that are modelled, take on a much more physical quality, especially when materials are added and the scene is rendered. Rendering is a process by which the computer adds virtual textures and features to the model's faces to give a more realistic finished product.

Another application for the use of 3D Studio Max and Cult 3D software could be in the production of interactive virtual learning manuals. These manuals would allow users to learn simple assembly tasks, using interactive and animation elements to demonstrate how the assembly is constructed. Training by this technique removes the danger element that would exist while experience is being gained in real life situations; for example the training of operators and maintenance technicians on large and potentially dangerous machines. It can also be used for training personnel for complex assembly processes.

1.6 Design and concurrent engineering

Concurrent engineering is no longer a new concept. A good number of years have passed since its value was identified in the 1970s. However, there are many opportunities to put concurrent engineering into practice. The Internet and 'real-time' communications have enabled many engineering companies to realise the potential of concurrent engineering techniques. Manufacturing systems have been designed to allow operations to run in parallel; design work, in particular, is conducive to these methods. Designing and prototyping can be undertaken together and often on an international basis. The following example shows how this can be achieved.

1.6.1 *Multidiscipline in the aircraft industry*

In the early days of the aircraft industry, aeroplanes were much simpler and this enabled engineers to be multidisciplined. As aircraft became progressively more complex, design engineers had to become specialised in narrower fields. However, as the power of computers has increased this has helped to reduce the level of complexity in the design process and has allowed engineers to become, once again, multidisciplinary. For example:

- products can be handled as 3D computer models from design inception to the simulation of cutting metal;
- tools can be designed allowing the engineer to construct structural analysis, determine the cost and producibility of a part, and simulate fabrication and assembly at the click of a mouse button.

Because computers can be networked across the world, this makes it possible for engineers to work in teams that are geographically dispersed. This applies both regionally and globally. For example the European Airbus is a collaborative project involving engineers in the UK and in France working closely together. The technology lets designers and engineers work as though they were next to each other, although they may be at the other side of the world.

The other main benefit of shared information is that a database of designs is available that can be re-used, rather than the engineer having to re-invent the designs. This allows engineers to be innovative when required, i.e. to increase their focus on more difficult problems.

It is not only the design engineer who may wish to use 3D virtual objects. Assembly line workers may find a 3D virtual interface extremely useful. For example assembly instructions are held as part diagrams and process control sheets. The opportunity to be able to interrogate a virtual 3D object on screen helps the assembly operator on the shop floor to understand the assembly process and overcome assembly problems more easily and with greater understanding.

1.7 Engineering successful product designs

No matter what design process is employed, a successful product requires the designer to strike a balance between price, product design, manufacturing and performance. Each

Fig. 1.26 *Product parameters*

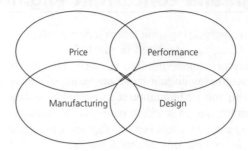

product has a different set of design parameters, goals, and specifications that are established early in the concept stage. For example the technical and economic objectives of components for a low-volume high-performance sports or racing car are significantly different from those for a high-volume family saloon.

Whatever the product, finding the correct balance has always been the biggest challenge faced by design teams. Maintaining this balance has gained even greater importance as manufacturing industry has had to become more competitive, and designers are forced to look for new and innovative design solutions to the age-old problem of how to do more with less. The days of over-engineering components and buying more capacity than is required have now been replaced with lean and common-sense manufacturing. For example, in this atmosphere of change, processes such as *turret punching* (Section 4.5) and *tube hydroforming* (Section 4.6) have proved themselves to be low-cost, highly efficient methods of production.

High-volume production is most sensitive to piece price and ease of manufacture but is less sensitive to tooling and investment costs as these can be amortised over a very large number of parts. This allows the use of high-speed dedicated production equipment. While product performance must meet the intended targets, it is often compromised in order to meet cost or manufacturing objectives.

Lower-volume products are more sensitive to total tool and investment costs. For this reason, low-volume parts are best produced using flexible machines that offer quick changeover times and use inexpensive tooling. The price of the final product piece, while always important, is of less significance when dealing in low-volume products. In this instance, expectations of product performance are often quite high, and this can justify a high piece price charged to the customer.

The four product parameters that must be balanced during product development are shown in Fig. 1.26. These are product price, performance, manufacture and product design.

The importance of these parameters will now be considered in greater detail.

1.7.1 *Product price*

Manufacturing industry breaks the price of a product into two areas, piece price and tooling. The required piece price or selling price for a part is the summation of direct manufacturing costs, indirect manufacturing costs and profit. Tool cost refers to any dies and fixtures that are custom made for the sole purpose of manufacturing the part. Tooling costs are often borne by the customer and, as such, do not directly affect the selling or piece price. Nonetheless, reductions in tool costs will often be considered in the sourcing decision.

1.7.2 *Direct costs*

Direct (variable) costs are those elements of manufacturing costs that can be directly attributed to the manufacture of the product. These costs are incurred in proportion to the amount of the product, they are not incurred if production is stopped. Some examples of direct costs are materials, consumables and direct labour. Let us now consider the effects of these costs.

Materials costs
Material is the biggest single direct cost of high-volume production. Efforts to reduce the materials cost can have dramatic effects on the final cost of manufacture.

Direct labour costs
Direct labour costs are largely driven by the degree to which the manufacturing process is automated, by the number of process steps, and by the remunerative expectations of the workforce.

Consumables
Consumables are items that must be replenished on a regular basis during production. These can take the form of lubricants, dies, mandrels, seal elements, punches, rust inhibitors, torch tips, weld wire, weld gases, electricity, process cooling water and shear blades. Minimising or eliminating the use of any of these items will reduce production costs.

1.7.3 *Indirect costs*

Indirect (fixed) manufacturing costs are those that cannot be directly attributed to the production of any given component. These costs will occur whatever the level of production is. Some examples of fixed costs are indirect labour, taxation, lighting, heating and depreciation. Except for *depreciation*, all of these examples of indirect costs are present regardless of how a product is designed or the methods used to produce it.

Capital plant, which has associated tooling, wears out in time or may be superseded by technological innovation and become obsolete. Therefore the value of such plant and tooling *depreciates* over a period of time. Thus, money must be put aside on a regular basis for the eventual replacement of the plant and equipment, and charged against the parts being produced. Cost control is dealt with in greater detail in *Manufacturing Technology*, Volume 2, by this book's authors.

1.7.4 *Performance*

All parts must meet the performance goals set out by the product's structures-engineering team. All products are put through rigorous validation testing during product development to ensure that each part will stand up to normal useage throughout the life of the product. The use of production-representative-prototype parts manufactured using production-intent processes during product development is key to ensuring that the actual production components will perform as intended. There are many different characteristics that make up the way in which structural performance is measured.

1.7.5 *Manufacture*

Ease of manufacture takes on much more importance when dealing with high-volume applications. A process that produces excessive scrap, products that must be reworked, or components that must be closely inspected to ensure quality will quickly overwhelm the plant and have excessive costs. The use of production-representative-prototype processes will quickly determine the viability of the manufacturing method chosen to produce a designed item. When problems are discovered during prototyping, it is important to address them using techniques that can be carried through into production, and are not just laboratory solutions. When taking a manufacturing issue, the first course of action should be to look for a design solution. Making a small modification to the product design to eliminate a concern is always preferable to patching a problem with increased operations, or to tightening specifications to guard against problems in production.

- *Process robustness/fault tolerance*. This, in broad terms, is the ability of the manufacturing process to absorb changes in inputs without affecting the final output. For example the process must be able to accept or, at the very least, adjust quickly to the inevitable differences between material batches. In production it is not possible to check every material batch in advance and choose only the ones that are most satisfactory.
- *Manufacturing scrap*. The reduction or elimination of manufacturing scrap provides two benefits. The first and most obvious is the reduction or elimination of material waste. The second is an increase in the amount of useable product-throughput. With a reduction of manufacturing scrap, additional equipment time is gained in making saleable products, since the production line is no longer making waste.
- *Number of operations*. Reducing the number of operations will offer the benefits of lower labour costs, reduced maintenance, less floor space and savings in equipment purchase costs.
- *Engineered scrap*. Reducing or eliminating engineered scrap will reduce material and scrap handling costs. The elimination of engineered scrap on a part will have the additional benefit of eliminating any additional corrective process, and its related capital, tooling and consumable costs.

1.7.6 *Product design*

All the factors mentioned above – price, manufacturing considerations and performance – come into play in the design of the product. The product design is the embodiment of all trade-offs. Every component has different requirements that have to be addressed during the product design phase. Different load patterns, service requirements or project goals can result in substantial differences in the design of the same part for its use on different products.

Section 1.7 is based on data provided by Mr Derek Payne of VARI-FORM Inc., Ontario, Canada.

1. Research the different Web sites that relate to 3D modelling packages. Analyse the different types, and the strengths and weaknesses of each. Some example sites are:
   - www.kinetix.com
   - www.solidedge.com
   - www.rhino.com
   - www.amapi.com

2. Research the different Web sites that relate to Web-based interactive programming models. Analyse the different methods (including programming methods) each one uses. Some example sites are:
   - www.cycore.com
   - www.macromedia.com (look for Director 8.5)

3. Using the virtual object shown in Chapter 1 select your own object, for example a cassette player, compact disc (CD) player or minidisk player, and model it using the same method.

4. Using Cult 3D, follow the example shown in Chapter 1 and make your own object truly interactive.

5. Research different ideas of how virtual objects may be used, for example simulated aircraft maintenance training, medical operations, etc. Write an essay of 1000 words that outlines how each use may be achieved and its benefits.

6. Imagine being the head of a company that manufactures motor cars. Give examples of how concurrent engineering can help the company work more effectively and efficiently.

7. Using the cassette player example shown in Chapter 1, give examples of how the cost of the product can be broken down into its final sale price.

2 Rapid prototyping techniques

When you have read this chapter you should be able to:

- appreciate the variety and applications of rapid prototyping techniques;
- understand the benefits of rapid prototyping techniques;
- discuss material addition techniques;
- discuss material subtraction techniques;
- appreciate the advantages and techniques of reverse engineering;
- analyse a worked case study.

2.1 Introduction

In the previous chapter we saw how the Internet could be used to market and test products. However, further physical testing of the product must take place in order to confirm tactile properties, ergonomics and other human-to-product interface issues. One technology that assists in this area is rapid prototyping (RP). By compressing development times RP makes it possible to respond to the demands of niche markets and to the more frequent introduction of new products. Once a company decides that it is going to pursue the RP route the next decision that will have to be made is the choice of the RP method to be adopted. This will depend upon the quantities involved, the complexity and the application of the RP model and the purpose for which the models will be used. For example:

- patterns for castings;
- electrodes for electrodischarge machining (EDM) of dies;
- marketing models.

When the models are used to produce castings in the finished material it is important to remember that they can only be used for limited testing and to see if the parts fit together. The previous chapter showed the many advantages of virtual products. However, the disadvantage is in being unable to physically grasp the product. This is no problem as the geometry used in a virtual design can be used to produce the finished physical item or a rapid prototyped component, so let us now consider the technique of *rapid prototyping*.

2.2 Rapid prototyping

As we saw in the previous chapter, data can be transferred from one system to another. This is achieved by exporting data from one application and importing this data into another application. The ability to export and import files from one system to another has led to large-scale system integration. Modern industry and working practices have adopted such system integration to increase efficiency by, for example, using common data.

Rapid prototyping is a modern technique that uses this approach. For example, geometry imported from virtual modelling, as described in the previous chapter, or directly from 3D computer aided design (CAD) geometry can be used to produce solid objects by a process known as rapid prototyping. This has the following advantages:

- It allows designers to view their work very quickly.
- There is no need for fully dimensioned detailed drawings.
- The digital data can be transmitted by e-mail directly to a company specialising in rapid prototyping.

2.2.1 *The advantages of rapid prototyping*

There are, obviously, many advantages in being able to produce a physical model quickly and relatively cheaply that can be handled by both the designer and the client. For example it:

- produces visual models for market research, publicity, packaging, and so on;
- reduces 'time to market' for a new product;
- generates customer goodwill through improved quality;
- expands the product range;
- reduces the cost and fear of failure;
- improves design communication and helps eliminate design mistakes;
- converts 3D CAD images into accurate physical models at a fraction of the cost of traditional methods, because there are no tooling costs;
- can be used as a powerful marketing tool, because the actual prototype rather than an illustration can be seen.

2.2.2 *Further applications of rapid prototyping*

The ability to produce components of extremely varied shapes and the inclusion of cavities has enabled many other areas of design, science and medicine to use this technique, for example:

1. Medical applications:
   (a) organic knee joints;
   (b) pre-surgery models taken from computer aided tomography (CAT) scans, which are used to plan the best operating strategy.
2. Scientific applications:
   (a) scale models of chemical and biological structures, at molecular level;
   (b) raised maps of the earth, i.e. to display countries and their topography to scale.

3. Mathematical applications:
   (a) demonstrating scale models of mathematical functions and wave forms.

2.3 Input data for rapid prototyping

Rapid prototyping-machine software deals with geometry by importing *sterolithography* (STL) files. The *sterolithography* format is an ASCII or binary file used in manufacturing. It is a list of the triangular surfaces that describe a computer generated solid model. This is the standard input for most rapid prototyping machines and is one of the basic file types that such packages use either to rapid prototype an object using desktop milling or for some other rapid prototyping technique such as *sterolithography apparatus* (SLA).

The CAD data is processed by slicing the computer model into layers, each layer being typically 0.1–0.25 mm thick. The machine then uses this data to construct the model layer by layer, each layer being bonded to the previous one until a solid object is formed. Due to this laminated method of construction a stepped surface is developed on curved faces, as shown in Fig. 2.1. The removal of this stepped surface is essential if maximum advantage of the process is to be realised.

There are two main rapid prototyping techniques:

1. Material addition techniques that build up the physical model in layers. There are a number of techniques such as:
   (a) Sterolithography Apparatus;
   (b) Selective Laser Sintering (SLS);
   (c) Solid Ground Curing (SGC);
   (d) Laminated Object Manufacturing (LOM);
   (e) Fused Deposition Modelling (FDM);
   (f) 3D printing.
2. Material subtraction processes such as 'desk-top' milling that remove waste material from a solid block to produce the desired physical model.

Let us now consider these processes in more detail.

Fig. 2.1 *Stepped construction*

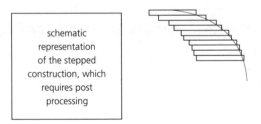

schematic
representation
of the stepped
construction, which
requires post
processing

2.4 Sterolithography Apparatus

The STL file of the component is sliced by the device's software. Each slice is then etched onto the surface of a photosensitive ultraviolet-curable resin using a 'swinging'

Fig. 2.2 *Schematic diagram of the sterolithography (SLA) apparatus*

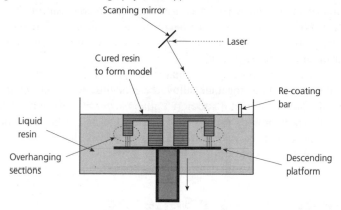

Fig. 2.3 *Adaptation of the SLA process by adding a snorkel*

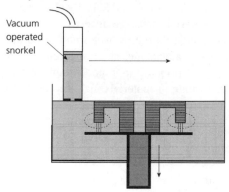

laser. Where the laser beam strikes the surface, the resin is cured. Each layer is typically 0.13 mm thick. At the end of each pass, which covers the whole of the surface of that layer, the platform descends to allow liquid resin to flow over the previously cured layer. A re-coating bar passes over the surface to ensure that a consistent layer thickness is achieved. This blade ensures that no air is trapped between the layers.

Models that have an overhang are produced on 'stilts'. That is, the software anticipates these problems and produces a platform on which the model is produced. To avoid the model sticking to the build platform, a lattice base is automatically created thus reducing the contact area of resin to platen. This technique is shown in Fig. 2.2. On completion, the model is carefully removed and washed in a solvent to remove uncured resin and placed in an ultraviolet oven to ensure all resin is cured.

An improvement to this technique has been the development of a *snorkel* as shown in Fig. 2.3. This is filled at the start and end of each pass. As the snorkel passes over the model, a controlled amount of resin is released ensuring that the entire surface is evenly coated. This technique ensures that resin cavities are not formed in areas that were not cured by the laser. For example, internal cavities that were filled with uncured resin are liable to destroy the model because of the internal expansion of the uncured resin upon curing. This method also allows different resins of varying viscosity to be used.

More recent developments have been prompted by problems caused by the expansion of the model where it is used as a disposable pattern (like the wax pattern in the Lost Wax Process). Where the resin model is produced with solid walls, expansion during the 'burning out' stage weakens the ceramic shell, and can cause failure in the firing and/or casting stages.

An engineering company that specialises in SLA systems has developed a machine and software that, together, allow the model to be constructed in the form of a honeycomb. The honeycomb structure collapses in on itself during 'burning-out' thus avoiding the problems of expansion. Each pocket of the honeycomb structure is connected to its neighbour by a small hole and this allows the uncured resin to be drained out of the prototype model prior to use.

2.5 Selective Laser Sintering

Sterolithography Apparatus and Selective Laser Sintering processes are very similar, but use different materials and take different forms. Whereas SLA uses a liquid ultraviolet-curable resin, the SLS process uses a powdered material. The use of powdered material is one of its major advantages since, theoretically, any powder that can be fused is capable of being transformed into a model and can be used as a true prototype component. Currently, the range of materials includes the following:

- Nylon.
- Glass-filled nylon.
- Polycarbonate.
- Waxes.
- Metal – this requires both a post-sintering operation and copper infiltration cycle before the part can be used.
- Sand bonded with heat-cured resin, for example casting patterns, that can be used in reverse engineering.

As with other rapid prototyping systems, the laser scans each slice of the component. In the case of SLS the powder is sintered by the action of the laser scan. As each layer is sintered, the platform lowers the cured layer of powder and a roller spreads a further layer of powdered material over the previous layer. This is shown in Fig. 2.4. During the build cycle any non-sintered powder helps to support the model, thus avoiding the need for stilt supports as with the liquid-resin system of the SLA process. When the prototype is removed from the bath of powder it requires a light brushing to remove any loosely attached material.

2.6 Solid Ground Curing

Solid Ground Curing is a physical imaging technology used to produce accurate, durable prototypes. Models are built in a solid environment, eliminating curling, warping, support structures and any need for final curing. The process consists of the following:

Fig. 2.4 *Schematic diagram of the selective laser sintering (SLS) process*

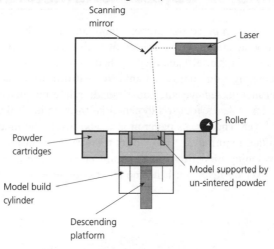

- A computer analysis of a CAD file that processes the object as a stack of 'slices'.
- The image of each working slice is 'printed' on a glass photo-mask using an electrostatic process similar to laser printing.
- That part of the 'slice' representing solid material remains transparent.
- This photo-mask is now used to produce the component slice by ultraviolet projection onto a photo-reactive polymer. The ultraviolet light cures the polymer resin.
- The unaffected resin (still liquid) is removed by vacuum.
- Liquid wax is spread across the work area as the process proceeds, filling the cavities previously occupied by the unexposed liquid polymer resin.
- A chilling plate hardens the wax, and the entire layer, wax and polymer, is now solid.
- The layer is milled level to remove any excess material.
- To build the next layer of the prototype, a further layer of photoreactive polymer is laid down on the previous work surface and spread evenly.
- An ultraviolet floodlight is again projected through the next photo-mask onto the newly spread layer of liquid polymer. Exposed resin again corresponding to the solid cross-section of the part within that slice polymerises and hardens.
- The process is repeated slice by slice, each layer adhering to the previous one, until the object is finished.
- The wax is removed by melting or rinsing, revealing the finished prototype (alternatively, it can be left in place for shipping or security purposes).

The Solid Ground Curing method has 10 to 15 times the throughput of other rapid prototyping methods based on photoreactive polymers. Any geometric shape can be created in any orientation. Parts can generally be made overnight, often in batches, and need no extra curing after they emerge. The use of wax means that supports do not have to be added to overhangs. Additionally, a build session can be interrupted to, for instance, expedite another project, and erroneous layers can be erased.

2.7 Laminated Object Manufacturing

As the name implies, the process laminates thin sheets of film (paper or plastic), the laser only has to cut the periphery of each layer and not, as in SLA, the whole surface.

The build material (paper with a thermo-setting resin glue on its underside) is stretched from a supply roller across an anvil or platform to a take-up roller on the other side. A heated roller passes over the paper, bonding it to the platform or previous layer, as shown in Fig. 2.5. A laser, focused to penetrate through one thickness of paper, cuts the profile of that layer. The excess paper around and inside the model is etched into small squares to facilitate its removal. Prior to its removal, this surplus material provides support for the developing model during the build process. The process of gluing and cutting continues layer by layer until the model is complete, as shown in Fig. 2.6.

Fig. 2.5 *Schematic diagram of the laminated object manufacturing (LOM) process*

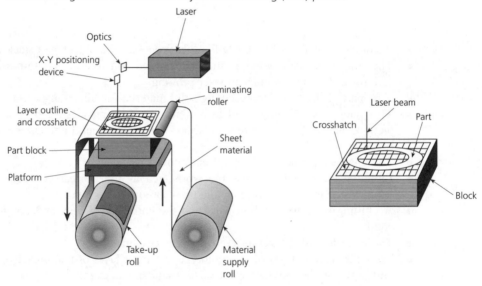

Fig. 2.6 *Part build up using laminated object manufacturing (LOM)*

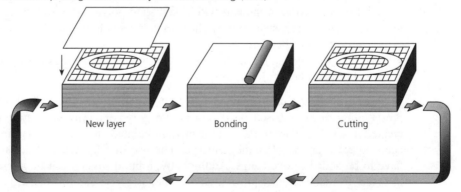

To reduce the build time, double or even triple layers are cut at one time, which increases the size of the steps on curved surfaces and the post-processing necessary to smooth those surfaces. Laminated Object Manufacturing objects are durable, multi-layered structures that can be machined, sanded, polished, coated and painted. They can be used as precise patterns for secondary tooling processes such as rubber moulding, sand casting and direct-investment casting.

2.8 Fused Deposition Modelling

The materials suitable for this process include the following:

- Acrylic-Butadiene-Styrene (ABS), also known as high impact polystyrene.
- Medical ABS.
- Investment casting wax.
- Low- and high-density polyethylene.
- Polypropylene.

A thermo-polymer is extruded from a travelling head having a single, fine nozzle. The head travels in the X-axis while the table or platform travels in the Y-axis and descends at predetermined increments in the Z-axis. On leaving the nozzle the thermo-polymer adheres and hardens to the previous layer, as shown in Fig. 2.7.

Fig. 2.7 *Schematic diagram of the fused deposition modelling (FDM) process*

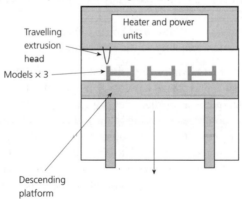

2.9 Three-dimensional printers

Whereas an ordinary printer lays down a single 2D layer of ink on a sheet of paper, these new devices can deposit a variety of materials. They add the extra dimension simply by printing layer after layer until there is a solid 3D object. This technology is maturing fast, and 3D printing is already used worldwide by engineers in many situations, for example:

Fig. 2.8 *Process stages in three-dimensional printing*

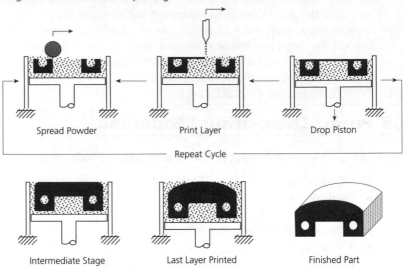

- plastic prototypes;
- the US army is currently developing a method of recreating engine parts;
- the US National Aeronautics and Space Administration has been successfully testing this technology in space;
- archaeologists have successfully recreated skulls of Egyptian mummies without removing the bandages;
- the recreation of dinosaur bones, where the data is sent from the field and the bone is recreated remotely in the museum.

The familiar 2D printer deposits ink, but the 3D version lays down droplets of hot, liquid plastic. The plastic hardens as it cools, building the desired pattern, layer by layer. This technique was further developed at the Massachusetts Institute of Technology (MIT) and combines sintering, gluing and droplet technologies. The process sprays fine droplets of glue onto specific points on a powder bed. This sticks the powder together in the right places, see Fig. 2.8. Then another layer of powder is deposited, rolled flat and glued. The finished product emerges when the layers are complete and the unglued powder – which provides support while the structure is growing – is blown away.

The disadvantage with all of these processes is that the object can only be made of a single material, metal, plastic or whatever. The real power will come when 3D printing can combine materials allowing us to create or recreate anything.

Current technology also allows ink-droplets to seep into the air gaps between the powder granules, giving colour right through the finished object. This is a step closer to inserting other materials into the voids to create a composite material. Once perfected, this could lead to a number of innovations, for example recreating printed circuit boards. A new motherboard could be printed directly from the Internet. Recent research has shown that features as small as 0.2 mm can be printed. There is also the ability to arrange the alignment of particles so as to increase strength in certain directions where it is required. Other features include cavities to save weight, depositing the material where strength is required and omitting particles where it is not required.

Recent analysis shows that a simple, low-cost 3D printer could be available soon and will be able to print out children's toys from data supplied over the Internet.

2.10 Material subtraction

This technique is in effect advanced computer numerical control (CNC) die milling, however, current software requires the same input data as the material addition techniques. Also parametric software reduces the skill required to generate the cutter paths. The main advantages of material subtraction are speed of production and strength of materials used.

However, the main disadvantages are that:

- it is limited to models with no cavities;
- no re-entrant angles can be machined, that is, the shape must be able to be machined with a vertical cutter. Although, if the shape curves underneath, the product can be turned over and the other side machined.

Most rapid prototyping systems are based on incremental-build techniques as in laminated manufacturing technology (LMT). Conversely new software that is currently available, such as DeskProto, is a *decremental* approach. This new system starts with a solid block of material (any material can be used) and removes as much as is needed to create the prototype. The decremental build technique fulfils the four criteria for rapid prototyping.

1. The software can import STL files, which are standard files for any rapid prototyping process.
2. The software builds using an automatic process and, in this, differs from other computer aided manufacturing (CAM) systems in that it is easy to use and functions almost automatically.
3. The software creates a prototype within a short timespan. The actual milling time with DeskProto can be very short indeed. Even more important is the short preparation time required, for example:
   (a) milling can be done without an operator;
   (b) translation from CAD geometry to CNC toolpaths is very quick.
4. Short delivery times, i.e. the complete system of software and milling machine is ideal for in-house use. The DeskProto company has tried to create a 'black box' approach, i.e. a highly automated system.

2.10.1 *Desk top milling using DeskProto*

Desktop milling is usually achieved by using small but extremely fast traverse CNC type machines. For example the Roland 3100 is very much like a 3D printer that plugs into the parallel port of a PC. The tool paths are generated directly from 3D geometry such as 3DS, DXF and STL files. The rest of the machining process is fully automated using simple, such as tooling and feed rate, parameters. The cutter paths are generated automatically and follow a given path direction, for example:

- parallel
- spiral
- crosswise

The main advantages of desktop milling over other RP techniques are as follows:

1. The prototype can be made from more robust material, i.e. aluminium.
2. Small series produced by (vacuum) casting in silicone-rubber moulds, created from a milled master prototype.
3. Milled moulds in tooling board for vacuum forming and hand lay-up in polyester. Two typical users in this field are:
   (a) chocolate moulds;
   (b) blister packs.
4. Milled tools in special tooling board for sheet metal forming. This material can be easily polished after milling and is ready for use in the sheet metal press.
5. Models in foam or wax for lost foam/wax metal casting processes, for example:
   (a) aluminium using lost foam;
   (b) gold and silver jewellery using lost wax.
6. Milled aluminium tools (moulds) for both cavity and core of small series injection moulded parts. This is extremely useful for small runs of injection moulded parts. The aluminium moulds will not last as long as a steel mould but are very useful for a company wishing to test the market with small run of products, i.e. conducting a market survey.
7. Electrodes for spark erosion – to be used to create production tools in steel.

2.11 Reverse engineering

Reverse engineering is often mentioned in rapid prototyping. This is where an engineering geometry is created by accurately measuring an existing physical product. The 3D scanners, which are used to measure an existing product, produce a cloud of point data. It is quite difficult to convert such data into a valid CAD model consisting of Non Uniform Rational B-Splines (NURBS) surfaces and/or perfect solids. However, packages such as DeskProto can convert the cloud point data into an STL file. Techniques that are able to probe or scan 3D objects are now discussed.

2.11.1 *Probing*

Coordinate measuring machines have evolved, from origins founded in simple layout machines and manually operated systems, into highly accurate automated inspection centres. A major factor in this evolution has been the touch trigger and other forms of inspection probes, and subsequent innovations such as the *motorised probe head* and the *automatic probe exchange system* for unmanned flexible inspection. The Renishaw touch trigger probe was originally developed in conjunction with Rolls-Royce plc, when a unique solution was required that would allow accurate pipe measurement for the

Anglo–French Concorde aircraft's engines. The result was the development of the first *touch trigger probe*: a 3D sensor capable of rapid, accurate inspection with low trigger force. Touch trigger probes are widely used for toolsetting and for in-cycle inspection. They can also be used to determine 3D geometrical points in space.

2.11.2 *Scanning*

Scanning is the term used to describe the process of gathering information about an undefined three-dimensional surface. It is used in such diverse fields as tool and die making, mould making, press-tool making, aerospace, jewellery, medical appliances and confectionery moulds. It is used whenever there is a need to reproduce a complex free-form shape. During the scanning process an *analogue scanning probe* (**not** a touch trigger probe) is commanded to contact and move back and forth across the unknown surface. During this process the system records information about the surface in the form of numerical data. This data is then used to create a CNC program that will machine a replica or geometric variant of the shape. Alternatively the data can be exported in various formats to a CAD/CAM system for further processing. This is shown in Fig. 2.9.

Fig. 2.9 *The scanning process: (a) scan – a range of devices capture data from an initial part/model; (b) manipulate data – create variant of original; (c) create output – CNC program or export data to CAD system; (d) replica; (e) rough machine mould (CNC mill); (f) finish mould cavity by CNC-type EDM; (g) CAD display if required*

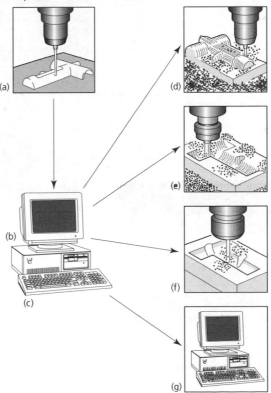

Fig. 2.10 *Scanning with an analogue probe: (a) scanning; (b) computer display; (c) CNC machining of replica*

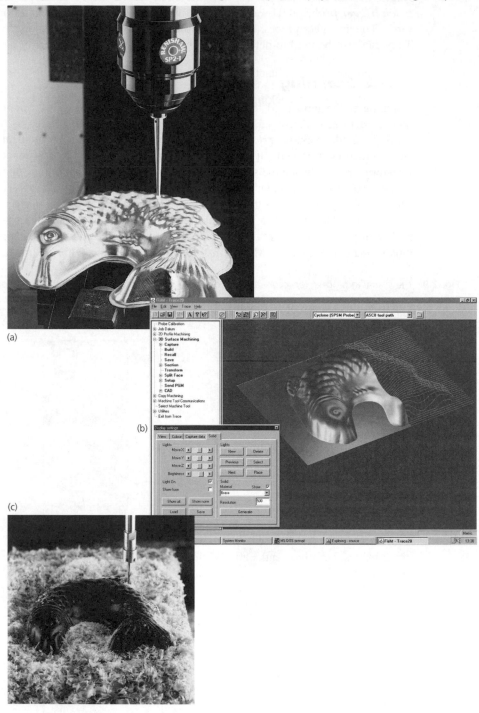

Let us look briefly at the difference between digitising and scanning. Digitising refers to the process of taking discrete points from a surface with a touch trigger probe. This time consuming process has since been greatly improved by the introduction of scanning. The scanning process uses analogue probes that produce a stream of data at up to 1000 points per second.

In addition to the contact probes so far discussed there also are:

- video probes for checking two dimensional forms such as circuit boards, holes and other features;
- optical probes using visible laser light to scan 3D objects made from pliable or delicate materials. The scanning takes place without deflection of the material.

Although surfaces can be scanned using video and optical laser probes, where the object is made from a rigid material the use of contact probe systems has several fundamental advantages:

- treatment of surfaces to prevent reflections is not required;
- vertical faces can be accurately scanned;
- data density is not fixed and is automatically controlled by the shape of the component;
- time-consuming manual editing of data to remove stray points is not required.

Figure 2.10(a) shows a three-dimensional shape being scanned. The data gathered by the probe is processed by the computer software package as shown in Fig. 2.10(b). The processed data can then be used to program a CNC machine so that it will make a replica of the original object, as shown in Fig. 2.10(c). Alternatively, the processed data can be used to make a mould for mass-producing replicas of the original object.

2.12 Case study

The following case study shows how data can be imported using STL format, then processed using DeskProto and final machined using a Roland CAMM 3 PNC 3100 modelling machine.

The cassette player from Chapter 1 was chosen as the subject for this case study. The method of creating geometry is exactly the same as that described in Chapter 1. Using 3D Studio Max an exact solid model was created. The mini CD player was split into 2 halves, i.e. a base and a lid. All single elements of the lid and the base were made into one object using the 'Attach' command.

The 'Lid' and 'Body' files were then exported as. STL files for transfer into DeskProto. Figures 2.11(a) and 2.11(b) show the base and lid imported into DeskProto.

The first task is to ensure that the model is capable of being machined on the computer-controlled machine. This is done automatically within DeskProto by specifying the machine that will be used. An automatic warning tells the user if the machine limits are exceeded. In the case study this happened and the part had to be scaled down by a factor of 0.62 using the 'transform' function within DeskProto.

Next, the cutter paths are automatically generated by clicking the 'calculate tool paths' button. The following parameters had been automatically set prior to this:

Fig. 2.11 *Case study: (a) lid; (b) body*

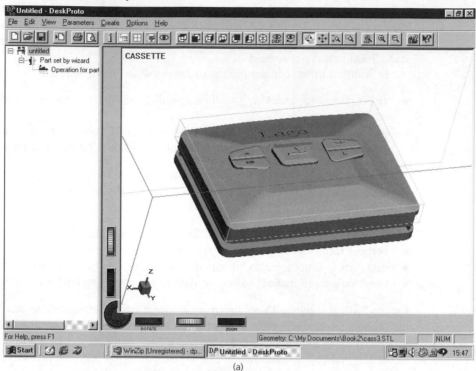

(a)

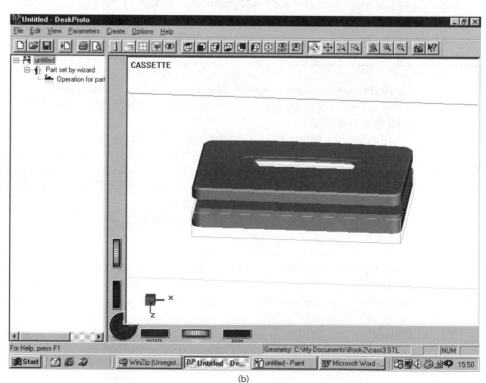

(b)

Fig. 2.12 *Simulated cutter paths: (a) lid cutter paths; (b) base cutter paths*

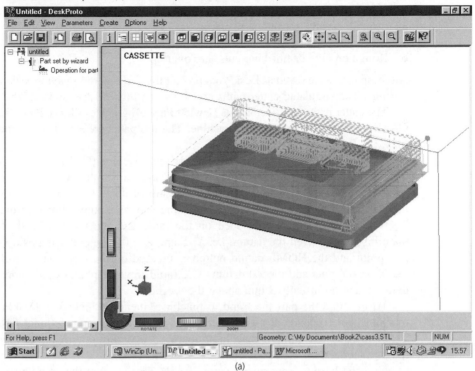

(a)

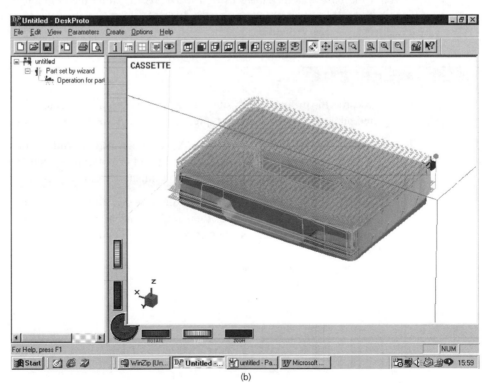

(b)

- Cutter size, in this case a 6 mm ball end cutter.
- Feed rates, in this case 16 mm per second, both horizontally and vertically.
- Spindle speeds, set by the operator.
- Machining strategy, i.e. horizontal, vertical weave or spiral.
- Roughing cuts or finishing cuts, i.e. overlap.

All these files were saved as DeskProto project files, in order to save the scale and machine settings. The simulated cutter paths are shown in Figs 2.12(a) and 2.12(b).

The cutter paths are generated as Hewlett Packard graphics files (HPGL) for the Roland CAMM 3 PNC 3100 modelling machine. The toolpaths are saved as Roland machining files, i.e. BASE.rol and LID.rol.

Using the Roland computer-aided modelling machine (CAMM) is very easy. It acts as a large 3D plotter and uses a parallel port connector to the PC. It acts very much like a printer, i.e. it can handle a continuous stream of data, and there are no limits to the file size it can handle, in very much the same way as a conventional printer.

A block of foam is positioned on the table and the top left-hand corner of the material is considered the datum, i.e. X, Y and $Z = 0$. The cutter is positioned to touch this point and the HOME datum button is pressed on the machine controller for both the X and Y data and the Z0 datum. The rapid traverse planes of Z1 and Z2 are also defined, and are usually 3 mm above the workpiece.

To machine the part the 'send to machine' menu is clicked in DeskProto and the machine tool responds at a rate four times faster than a conventional CNC machine. This is because the cutter is machining foam, resin or aluminium, not die steel nor other tough materials.

First the base was machined using the BASE file, then the component was turned over and the lid was machined using the LID file. In this way a complete model of a cassette player was produced.

ASSIGNMENTS

1. Research the different Web sites on rapid prototyping and determine the cost and time taken to rapid prototype typical products. Also state size limitations.

2. Research the different Web sites on CAD packages and determine which ones would be the most useful in exporting data to a rapid prototyping system. Determine the file format and if a representation of the finished product is given.

3. Visit the www.deskproto.com Web site and download a free trial copy of the software. Install this software, then import one of your own CAD models and use the interactive menu system to simulate the cutter paths that would machine the object.

Part B
E-manufacture

3 Advances in manufacturing technology (1)

When you have read this chapter you should be able to:

- understand the basic principles of CNC part programming;
- understand the principles of more advanced part programming;
- appreciate the need for advanced manufacturing technology;
- appreciate the advantages and limitations of advanced manufacturing technology.

3.1 The need for advanced manufacturing technology

The Department of Trade and Industry (DTI) has defined advanced manufacturing technology (AMT) *as the application of computers to manufacturing operations*. The importance of AMT cannot be overstated. For example, when the Advisory Council for Applied Research and Development (ACAR) published its report on *New Opportunities in Manufacturing* in 1984 it stated that, 'companies not adopting AMT may not exist in 10 years time'. The main elements of AMT as defined by the DTI will be covered in this and the following chapters. These elements are:

- computer numerical control;
- computer aided design and manufacture;
- flexible manufacturing systems (FMSs) – see Section 5.10;
- computerised management systems, i.e. inventory control, production control, and automated stores and parts issue – see Chapter 6.

Figure 3.1 shows the inter-relationship between these elements so that they become a wholly integrated system.

In recent years the manufacturing sector of the UK has been under intense pressure owing to a number of factors that threaten the very existence of many companies. These factors are:

- increased market competition from local and overseas competitors;
- shorter product life cycles;

Fig. 3.1 *Areas of advanced manufacturing technology (AMT)*

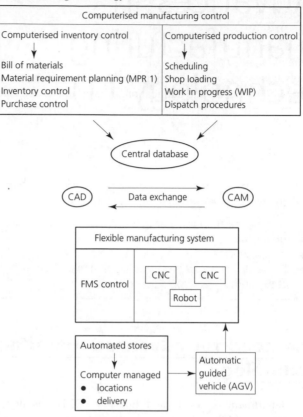

- shorter working week and increasing labour costs;
- high cost of capital investment in new plant and equipment;
- unpredictable economic situations;
- greater sophistication of consumers who demand higher quality, reliability and variety of products;
- pressure to produce goods within tighter environmental constraints that, in turn, add to the cost of the product and perhaps the use of more sophisticated processes.

Let us now look at what a company must do to overcome these factors in order to survive:

- reduce the lead time from the design inception stage to producing the finished product;
- increase the flexibility of the manufacturing process both in terms of handling part variety to accepting design alterations and scheduling changes;
- increase response to changes in the market demand;
- reduce the product costs;
- increase its overall efficiency.

Where companies have accepted and exploited the introduction of computers into industry they have been able to meet some of the survival criteria. The two most important computer applications in industry are Computer Integrated Manufacture (CIM) and Advanced Manufacturing Technology. We have already defined AMT, and we can now define CIM *as the unimpeded flow of electronic data within industry.* These computer applications are mainly concerned with increasing the flexibility and control of the manufacturing environment, while reducing the duplication of information and effort. This chapter is concerned with the application of computer numerical control to a number of manufacturing processes not only in the engineering sector but also in other sectors of manufacturing industry. Before considering CNC applications let us first look at the general principles of CNC programming as these are common across a wide range of machines and industries.

3.2 Computer numerical programming – an introduction

In a numerically controlled machine tool, the decisions that govern the operation of the machine are made by a series of numbers in binary code, which are interpreted by an electronic system. The electronic system converts these numerical commands into the physical movement of the machine elements. Thus each component is an exact replica of this stored data and high levels of repeatability and consistent quality are achieved.

Computer numerical controllers have built-in, dedicated computers. These have very powerful memory facilities that not only store sophisticated system management software, including frequently used standardised programming data (canned cycles), but also the whole part-programme. The system management software interprets the alphanumerical data of the part-programme and feeds the information for each operation into the electronic system that operates the machine movements.

Computer numerical control is used throughout industry. For example it is used in wood-machining, weaving and carpet-making, as well as in engineering. Some typical engineering applications are:

- Machining centres (milling machines)
- Turning centres (lathes)
- Drilling and boring machines
- Precision-grinding machines
- Centreless-grinding machines
- EDM (spark-erosion) machines
- Laser-cutting machines

Computer numerical control is also widely used for sheet metal working and fabrication. Typical applications of CNC to sheet metal and fabrication equipment are as follows:

- Turret punching machines
- Flame-cutting machines
- Forming machines
- Welding machines
- Tube bending machines

Computer numerical controlled, automatic inspection machines for checking two- and three-dimensionally contoured components are also available. This improves the frequency and accuracy of part inspection that is economical, with a corresponding improvement in quality control.

3.3 Computer numerical programming – the basic principles

Computer numerical programming is the process of converting the information on a component drawing into a format that the machine control software can recognise and act upon. The part program for any machine tool or system must be designed with *safety as the main priority*. Very large machine tools operating at high speeds and driven by powerful motors are potentially dangerous and must be treated with the utmost respect. For this reason the programs and programming techniques discussed throughout this chapter will follow the essential rules to ensure safe operation. Let us commence by considering a very simple program structure in which only one tool or cutter is used, and all the necessary cutting data can be loaded at the start. The structure of such a programme will contain the following information.

3.3.1 *Start-up information*

This is the first part of the program where any defaults such as English or metric units are set. It is also necessary to state whether positional information will be in an absolute or incremental format. A safe tool change position must be set with the associated tool offsets, speeds and feed values. Failure to associate speeds and feeds with the correct tool could result in the tool using the last defined (modal) speed and feed value stored in the machine register. This could be dangerous and result in injury and damage to the machine. Remember that *modal* commands continue to be active until they are cancelled or changed.

3.3.2 *Machining information*

This is the information that controls the relative movements of the cutter and/or the workpiece and thus generates the required shape and features. The workpiece may have a simple two-dimensional profile or it may have more complex three-dimensional contoured surfaces. The machining information must also contain such data as the depth of cut, macros and canned cycles for roughing and drilling, etc. As will be shown later, the cutter must always be introduced to the work on either a 'ramp' or a curved path so as not to leave a dwell-mark on the workpiece. This technique introduces the cutting forces gradually as the cutter is fed into its correct depth.

3.3.3 *Close-down information*

This is where the cutter returns to the tool change position away from the workpiece. At the same time the spindle drive is turned off and the program returns to the start position ready for the next component. This will allow the operator to unload or inspect the workpiece under safe conditions, away from any sharp cutters, and, if necessary, to change the cutting tool if the machine is configured for manual tool changing.

3.4 Computer numerical programming – the advantages and limitations

Although, at first sight, numerically controlled machines appear to be costly and complicated to use, the fact that they are now universally employed by industry is evidence that their advantages far outweigh their limitations.

3.4.1 *Advantages*

High productivity

Although there is no difference in cutting speeds and feed rates between CNC machines and manually operated machines of similar power and rigidity, production time is saved by rapid traversing and positioning between operations, and greater flexibility. For example, milling, drilling and boring can all be achieved on a CNC machine at the same setting. This avoids expensive jigs and fixtures; the amount of work in progress stored between operations, and the time spent in re-setting work as it is passed from one machine to the next. Computer numerical control machines do not become tired and maintain a constant, high rate of productivity. Their electronic computerised control systems allow them to be easily linked to a robot for work loading and unloading so that 'lights out' production can be maintained round the clock.

Tool life

Since the tool approach and cutting conditions are controlled by the programme and are constant from one component to the next, tool wear is more even and tool life is extended. Further, since profiles and contours are generated by the programme controlled workpiece and tool movements, complex and delicate form tools, which are susceptible to damage, are not required.

Work-holding

Since each workpiece merely has to be positioned relative to the same datum point on the machine table, as dictated by the programme, and securely clamped, complex jigs and fixtures are no longer required.

Component modification

When using jigs and fixtures on manually operated machines, even small changes to the workpiece design can result in costly modifications or even replacement of those jigs and fixtures. However, when using CNC machines it is usually only necessary to make small modifications to the part-program, taking but a few minutes, before production of the modified component can commence.

Design flexibility

Components with complex profiles and components requiring three-dimensional contouring can be more easily and accurately produced on CNC machine tools than on manually controlled machines. It is also possible to produce components on CNC machines that cannot be produced on manually operated machines.

Reduced lead time

The use of CNC reduces the lead time required to bring new components into production, compared with the use of conventional automatic machines, for the following reasons.

- Writing a part-program is very much quicker than producing the cams required for an automatic lathe.
- Complex form tools are not required since CNC machines generate the required profile from the program data.
- Workholding is simplified, so that complex jigs and fixtures do not have to be made.

Management control

Since machine performance is controlled by the program rather than by the operator, there is greater management control over the cost and quality of production when using CNC machines.

Quality

Because there is less operator involvement, CNC machines produce components of more consistent quality than manually operated machines. If the machine is fitted with *adaptive control* the machine will always run the tooling at the optimum production rate. It will also sense tool failures or other variations in performance and either stop the machine or, if fitted with automatic tool changing, select back-up tooling from the tool magazine. Automatic gauging can also be fitted.

3.4.2 *Limitations*

Capital cost

The initial cost of CNC machine tools is substantially higher than for similar manually operated machines. However, in recent years, the cost differential has narrowed.

Depreciation

As with all computer-based devices, CNC controllers rapidly become out of date. Therefore CNC machine tools should be 'written-off' over a relatively short period of time, and should be replaced more frequently than has been the practice with manually controlled machines.

Tooling costs

To exploit the production potential of CNC machine tools, specialised tooling has been developed for use with them. Although the initial cost is high, this largely reflects the tool shanks and tool holding devices that do not have to be frequently replaced. The cost of replacement tool inserts is no higher than for manually operated machines. The controlled cutting conditions also mean that insert replacement is less frequent while, at the same time, the tooling can be run continuously at its optimum performance levels.

Maintenance

Computer numerical control machine tools and their controllers are extremely complex and it is unlikely that small- and medium-sized companies will have the expertise to maintain and repair them, except for routine lubrication and adjustments. Therefore an approved maintenance contract for each machine or group of machines is required. Such contracts are expensive – usually about 10 per cent of the original purchase price per annum.

Training

Comprehensive programmer and operator training is required to convert the workforce from being expert in the use of conventional, manually operated machines to being expert in the programming and operation of CNC machines. Since there is little standardisation between different makes of controller or even between different types and generations of the same make, extensive type-specific familiarisation training is required for each new machine. Although familiarisation training is provided by the equipment supplier within the purchase cost of the equipment, it is time-consuming and therefore costly in terms of lost production.

3.5 Computer numerical programming – axis nomenclature

An important feature of the information that is supplied to the control system is slide displacement. Most machines have two or more slides (usually perpendicular to each other) and, in addition, these slides can be moved in one of two directions. It is therefore essential that the control system knows which slide is to be actuated, in which direction the slide is to move and how far the slide is to move.

British standard BS 3635 provides axis and motion nomenclature, which is intended to simplify programming and to standardise machine movements. The basic principle of this notation is shown in Fig. 3.2(a), from which it can be seen that there are three basic axes of movement.

- The Z-axis is always the main spindle axis and is positive in a direction towards the toolholder (away from the work). This is a safety feature, so that should the programmer omit the directional sign (in this case negative) from in front of the numerical positional data, the tool will always move away from the work.
- The X-axis is always horizontal and parallel to the working surface.
- The Y-axis is perpendicular to both the X- and Z-axes.

Once the positive Z direction has been found, the positive X and Y directions can be found from the 'right-hand rule' as shown in Fig. 3.2(b). Figure 3.3 shows examples of axis and motion nomenclature for typical machine tools in accordance with BS 3635.

Fig. 3.2 *Axis notation: (a) notation; (b) right-hand rule*

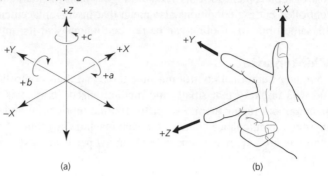

(a) (b)

Fig. 3.3 *Examples of axis notation: (a) axes for vertical milling machines and drilling machines; (b) axes for lathes and horizontal boring machines*

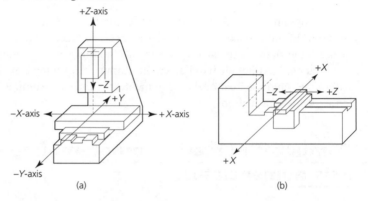

(a) (b)

The programmed movements of a CNC machine tool can be described in three ways.

- *Point-to-point systems.* These are designed to position the tool at a series of different points. The path of the tool between the points and the traverse rate between the points are under the control of neither the programmer nor the operator. Such a system is suitable for simple drilling operations or for a sheet metal turret-punching machine. It is not suitable for profile machining or contouring. A point-to-point system always takes the shortest path, so do not obstruct it.
- *Linear systems.* Also known as parallel path systems, they are designed to move the tool from one position to the next so that the tool path is a series of straight lines. Each line may be parallel to the X-axis or the Y-axis. The traverse rate is under the control of the programmer, and machining operations such as milling can take place between the tool positions. Such a system is unsuitable for profile machining or contouring.
- *Continuous path systems.* These are the most widely used systems. The path taken by the tool in moving from one point to the next and the traverse rate at which this occurs are fully under the control of the programmer. Angular and curved movements can be made in two or three axes simultaneously and complex profiles (two-dimensional) and contoured (three-dimensional) components can be generated.

For convenience when programming, it is always assumed that the cutter moves and follows the profile of the workpiece, despite the fact that, in practice, it is the worktable and work that moves. The computer in the controller automatically makes the translation from the programmed cutter movement to the actual work movement.

3.6 Computer numerical programming – control systems

3.6.1 *Open-loop control system*

In an open-loop control system the machine slides are displaced, according to information loaded from the part programme into the control system, without their positions being monitored. Hence there is no measurement of slide position and no feed-back signal for comparison with the input signal. The correct movement of the slide is entirely dependent upon the ability of the drive system to move the slide through the exact distance required. Such a system is shown in Fig. 3.4(a). The most common method of driving the lead screw is by a stepper motor either directly or via a toothed belt drive. A stepper motor is an electric motor energised by a train of electrical pulses rather than by a continuous electrical signal. Each pulse causes the motor to rotate through a small discrete angle. Thus the motor rotates in a series of steps according to the number of pulses supplied to it. The direction of rotation depends upon the polarity of the pulses and the feed rate depends upon the number of pulses per second.

3.6.2 *Closed-loop control system*

In this system a signal is sent back to the control unit, from a measuring device (called a transducer) attached to the slideways, indicating the actual movement and position of the slides. Until the slide arrives at the required position, the control unit continues to adjust its position until correct. Such a system is said to have feedback and is shown

Fig. 3.4 *Control systems: (a) open-loop control, no feedback; (b) closed loop control*

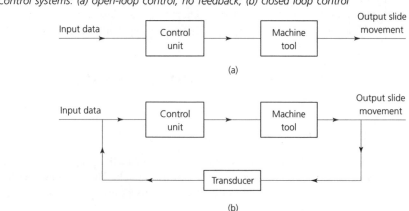

in Fig. 3.4(b). Although more complex and costly than open-loop systems, closed-loop systems give more accurate positioning especially on medium and large machines where the forces involved in moving the machine elements can be high. For this type of system servomotors are used to drive the leadscrews. Unlike stepper motors, these have a high starting torque to overcome the inertia of the machine slides and workpiece. They can be used on all sizes of machine up to the largest.

3.6.3 *Transducers and encoders*

Transducers are devices that convert one physical variable into another. For example a microphone converts sound waves into electrical signals. In CNC machines transducers may be used for a number of purposes, such as:

- monitoring the position of the worktable and, therefore, the work;
- measuring the speed and angular displacement of the machine spindle;
- measuring tool tip temperature;
- measuring cutting power being transmitted to the machine spindle;
- measuring oil pressure in the hydraulic and lubricating systems;
- measuring the volumetric flow of the coolant.

For example, in a closed-loop system linear transducers attached to the machine table provide the necessary feedback required for the servomotors to position the worktable and work accurately in accordance with the requirements of the program.

Rotary transducers are used to measure the angular displacement of rotating elements of a machine such as the leadscrew and spindle of a lathe. This is essential for synchronising the rotation of the work and the axial movement of the tool when screw-cutting on a CNC lathe.

It has already been stated that transducers provide incremental data (the distance moved from one position to the next). However, they do not know where they are relative to the machine datum. Also, they 'lose their memory' if the power supply to the machine fails or they are turned off. Therefore most systems use *encoders*. These are transducers using more complex scales to provide additional information so that the control system knows not only how far the worktable and work has moved, but also exactly where the work is from the machine datum. What is more, at switch on the machine immediately identifies the position of all the machine elements. Therefore encoders are able to provide both *incremental* and *absolute* information for the control system. Refer to Fig. 3.14 on page 69.

3.7 Computer numerical programming – data input

3.7.1 *Manual data input*

This can be used to enter complete programs, to edit the programs or to set the machine by manually pressing the keys on the control console. The manual loading of complete programs is rarely used and should be limited to simple programs so that machine idle time is kept to a minimum.

3.7.2 *Conversational data input*

The operator has again to load the program into the control console by manually depressing the appropriate keys. However, instead of writing out the program in machine code in advance of loading it, the program is entered in response to questions (prompts) appearing on a visual display unit (VDU) in *conversational* English. In this system the computer is pre-programmed with standard data stored in 'files' within the computer memory. Each item of data is numerically identified and called into the program by the response of the operator to a question. To reduce idle time, most modern conversational control units allow a new program to be entered while an existing program is still operating the machine.

3.7.3 *Punched tape*

Punched tape was the original method of loading data into numerically controlled machines. It is now virtually obsolete but may be found on some older machines that are still in use.

3.7.4 *Floppy disk and CD-ROM*

Because computer-aided programming software is now widely available and widely used, it is most convenient to record the program on computer disks. Such software has the advantage that the computer does most of the calculations and thus removes a major source of programming errors. Also the shape of the component is displayed on the screen, therefore the programmer can check the program before loading it into the machine. Furthermore, the programmer only has to be familiar with one 'language', the software will convert the program into the language of the machine controller. This is of great importance when a firm has many machines and controllers of different makes and languages.

3.7.5 *Direct numerical control*

The program is prepared on a remote computer and stored in the memory of that computer. It can be checked on the computer using simulation software as previously described. However, this time, the program is loaded 'down-line' directly into the CNC machine, with a copy of the program retained on disk. The computer software has to be compatible with the machine controller's management software.

3.8 Computer numerical programming – program terminology

3.8.1 *Characters and words*

A *character* is a number, letter or symbol that is recognised by the controller. A *word* is an associated group of characters that define one complete element of information, e.g. N150. There are two types of word: *dimensional* words and *management* words.

3.8.2 *Dimensional words*

These are any words related to a linear dimension. That is, any word commencing with the letters X, Y, Z, I, J, and K or any word in which the above characters are inferred. The letters X, Y, Z, refer to the corresponding machine axes as defined in Figs 3.2 and 3.3. The letters I, J, K, refer to circles and arcs of circles. The start and finish positions of arcs are defined by X-, Y-, Z-coordinates, while the centre of radius of the arc or circle is defined by I-, J-, K-coordinates with:

- I-dimensions related to X-dimensions;
- J-dimensions related to Y-dimensions;
- K-dimensions related to Z-dimensions.

Current practice favours the use of a decimal point in specifying dimensional words and a machine manual may stipulate that an X-axis dimension word has the form X4,3. That is, the X-dimension may have up to 4 digits in front of the decimal point and up to 3 digits behind the decimal point. This is standard practice when the dimensions are in millimetres. For dimensions in inches the form is X3,4. That is, up to 3 digits in front of the decimal point and up to 4 digits behind it.

In addition to stating the axis along which the machine element will move, and the distance it will move, the direction of movement must also be specified. To indicate the direction of movement the dimension is given a positive or negative sign. If there is no 'sign' in front of the digits, positive movement will be assumed. If there is a minus sign in front of the digits, negative movement is indicated. In the case of the Z-axis, the tool will move away from the work for safety if the minus sign is omitted as shown in Fig. 3.5.

3.8.3 *Management words*

These are any words that are not related to a dimension. That is any word commencing with the character N, G, F, S, T, M, or any word in which the above characters are implied. Here are some examples.

Fig. 3.5 *Z-axis movement*

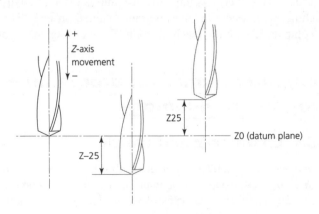

- *N4 is a block or sequence number*. The character N is followed by up to 4 digits (i.e. N1 to N9999). A block word is usually the first word that appears in any block and identifies that block. Blocks are usually numbered in steps of 5 or 10 so that additional blocks can be inserted if needed.
- *G2 are preparatory codes or functions*. The character G is followed by up to 2 digits (i.e. G0 to G99). These are used to *prepare* or inform the machine controller of the functions required for the next operation. The preparatory codes specified in BS 3635 are listed in Table 3.1. Unfortunately these codes vary slightly between different makes and types of controller, so the maker's programming manual should always be consulted. Many preparatory codes are *modal*. That is, once selected they remain in operation until changed or cancelled.
- *F4 are feedrate commands*. The character F is followed by up to 4 digits (i.e. F1 to F9999). They indicate to the controller the desired feedrate for machining and may be defined in terms of: millimetres per minute; inches per minute; millimetres per revolution; inches per revolution; or a feedrate number. Feedrate numbers are an older system in which typical feedrates in, say, millimetres per minute or millimetres per revolution are predetermined by the manufacturer and selected by an appropriate F-code, together with a G-code to tell the controller which system is being used.
- *S4 is a spindle speed command*. The character S is followed by up to four digits (i.e. 1 to 9999). Again there are various ways of defining the spindle speed, such as: revolutions per minute; cutting speed in metres per minute; and constant cutting speed in metres per minute. The latter method is used when facing across the end of a

Table 3.1 *Preparatory functions*

Code number	Function	
G00	Rapid positioning, point to point	(M)
G01	Positioning at controlled feed rate	(M)
G02	Circular interpolation — Normal dimensions	(M)
G03	Circular interpolation CCW*	(M)
G04	Dwell for programmed duration	
G05	Hold: cancelled by operator	
G06 G07	Reserved for future standardisation	
G08	Programmed slide acceleration	
G09	Programmed slide deceleration	
G10	Linear interpolation (long dimensions)	(M)
G11	Linear interpolation (short dimensions)	(M)
G12	3D interpolation	(M)
G13–G16	Axis selection	(M)
G17	XY plane selection	(M)
G18	ZX plane selection	(M)
G19	YZ plane selection	(M)
G20	Circular interpolation CW* (long dimensions)	(M)
G21	Circular interpolation CW (short dimensions)	(M)

Table 3.1 *(cont'd)*

Code number	Function	
G22	Coupled motion positive	
G23	Coupled motion negative	
G24	Reserved for future standardisation	
G25–G29	Available for individual use	
G30	Circular interpolation CCW (long dimensions)	(M)
G31	Circular interpolation CCW (short dimensions)	(M)
G32	Reserved for future standardisation	
G33	Thread cutting, constant lead	(M)
G34	Thread cutting, increasing lead	(M)
G35	Thread cutting, decreasing lead	(M)
G36–G39	Available for individual use	
G40	Cutter compensation, cancel	(M)
G41	Cutter compensation, left	(M)
G42	Cutter compensation, right	(M)
G43	Cutter compensation, positive	
G44	Cutter compensation, negative	
G45	Cutter compensation +/+	
G46	Cutter compensation +/–	
G47	Cutter compensation –/–	
G48	Cutter compensation –/+	
G49	Cutter compensation 0/+	
G50	Cutter compensation 0/–	
G51	Cutter compensation +/0	
G52	Cutter compensation –/0	
G53	Linear shift cancel	(M)
G54	Linear shift X	(M)
G55	Linear shift Y	(M)
G56	Linear shift Z	(M)
G57	Linear shift XY	(M)
G58	Linear shift XZ	(M)
G59	Linear shift YZ	(M)
G60	Positioning exact 1	(M)
G61	Positioning exact 2	(M)
G62	Positioning fast	(M)
G63	Tapping	
G64	Change of rate	
G65–G79	Reserved for future standardisation	
G80	Fixed cycle cancel	
G81–G89	Fixed cycles	(M)
G90–G99	Reserved for future standardisation	(M)

Notes: *CCW = counter-clockwise; CW = clockwise. Functions marked (M) are modal. These codes are based on BS 3635; but codes vary between makes and types of controller, and the manufacturer's programming manual should always be consulted. Most controllers use G90 to establish the program in *absolute* dimensional units, G91 to establish the program in *incremental* dimensional units. FANUK controllers use G20 in place of G90, and G21 in place of G91.

component so that, as the effective diameter gets smaller, the spindle speed is increased to a predetermined safe maximum so that the surface cutting speed is maintained at a constant value.

- *T2 is a tool number.* The character T is followed by up to 2 digits (usually 1 to 20) and identifies which tool is to be used. Each tool used will have its own tool number and the computer, as well as memorising the tool number, also memorises such additional data as the tool length offset and/or the tool diameter/radius compensation for each tool. In a machine with automatic tool changing, the position of the tool in the tool-magazine is also memorised by the computer under the tool number file.
- *M2 is a miscellaneous command.* The character M is followed by up to two digits (i.e. 1 to 99). Apart from the preparatory functions (G-codes), there are a number of other commands that are required throughout the programme. For example: starting and stopping the spindle; turning the coolant on or off; changing speed and changing tools. The miscellaneous codes are listed in Table 3.2.

Note: N, G, T, M commands may require a leading zero to be programmed on some older but still widely used systems. For example: G0 becomes G00; G1 becomes G01; M2 becomes M02.

An example of a CNC program using the above codes and commands could look like:

N5 G90 G71 G00 X35.4 Y25.5 T01 M06
N10 X15.0 Y15.0 S1250
N15 G01 Z–25.0 F120 M03

3.8.4 *Program format*

Different control systems use different formats for the assembly of each block of data. Thus the programming manual, for the machine being programmed, should always be consulted. A block of data consists of a complete line on a program containing a complete set of instructions for the controller. The *word* (or *letter*) *address system* is the most widely used system. Each 'word' commences with a letter character called an address. Hence a word is identified by its letter and not by its position in the block as in some older (*sequential format*) systems. The word (or letter) address system has the advantage that instructions that remain unchanged from a previous block may be omitted from the subsequent blocks until a change becomes necessary.

3.9 Computer numerical programming – basic part programming

3.9.1 Canned cycles

Standardised fixed cycles, or *canned cycles* as they are more commonly known, are used to save the repetitive programming of frequently used operations. The sequence of events for such a cycle of operations is embedded in the memory of the controller's computer

Table 3.2 *Miscellaneous functions*

Code number	Function
M00	Program stop
M01	Optional stop
M02	End of program
M03	Spindle on CW*
M04	Spindle on CCW*
M05	Spindle off
M06	Tool change
M07	Coolant 2 on
M08	Coolant 1 on
M09	Coolant off
M10	Clamp slide
M11	Unclamp slide
M12	Reserved for future standardisation
M13	Spindle on CW, coolant on
M14	Spindle on CCW, coolant on
M15	Motion in the positive direction
M16	Motion in the negative direction
M17 } M18 }	Reserved for future standardisation
M19	Oriented spindle stop
M20–M29	Available for individual use
M30	End of tape
M31	Interlock bypass
M32–M35	Constant cutting speed
M36	Feed range 1
M37	Feed range 2
M38	Spindle speed range 1
M39	Spindle speed range 2
M40–45	Gear changes
M46–49	Reserved for future standardisation
M50	Coolant 3 on
M51	Coolant 4 on
M52–54	Reserved for future standardisation
M55	Linear tool shift, position 1
M56	Linear tool shift, position 2
M57–M59	Reserved for future standardisation
M60	Workpiece change
M61	Linear workpiece shift, position 1
M62	Linear workpiece shift, position 2
M63–M67	Reserved for future standardisation
M68	Clamp workpiece
M69	Unclamp workpiece
M70	Reserved for future standardisation
M71	Angular workpiece shift, position 1
M72	Angular workpiece shift, position 2
M73–M77	Reserved for future standardisation
M78	Clamp slide
M79	Unclamp slide
M80–M99	Reserved for future standardisation

Notes: *CW = clockwise; CCW = counter-clockwise. N, G, T, M commands *may* require a leading zero to be programmed on some older but still widely used systems. For example, G0 becomes G00, G1 becomes G01, M2 becomes M02.

Fig. 3.6 *Canned drilling cycle (G81)*

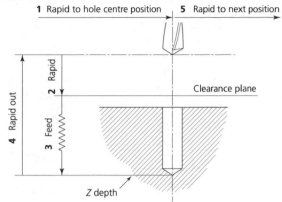

at the time of manufacture and is called up, when required, by an appropriate G-code. One of the most commonly used canned cycles is the drilling cycle that is shown in Fig. 3.6. This is called up by using a G81 code, and the cycle of events that the machine performs automatically is as follows:

- Rapid traverse to centre of first hole.
- Rapid traverse to clearance plane height.
- Feed to depth of hole.
- Rapid up to clearance plane height.
- Rapid traverse to centre of next hole – repeat for as many holes as required.

The only data the programmer has to provide is as follows:

- The positions of the hole centres.
- The spindle speed.
- The feedrate.
- The tool number.

The machine setter will provide the tool length offset for the drill and any other information that needs to be recorded under the tool number file in the computer memory.

Another important canned cycle used on milling machines is the rectangular pocket milling cycle, as shown in Fig. 3.7. This cycle is called up by using the G78 preparatory code. The movements involved in this cycle are as follows:

- Rapid traverse to position 1 inside the pocket boundary.
- Rapid down to the clearance plane height.
- Feed down in the Z-axis to the roughing depth (since the cutter is plunging into solid metal, a slot drill must be used and not an end mill).
- Machine out the pocket as the cutter path moves sequentially through positions 2 to 15 inclusive in Fig. 3.7 (less a finishing allowance on the profile).
- Rapid up to the clearance plane height in the Z-axis, and return to position 1.
- Feed down to finish pass depth.

Fig. 3.7 *Canned pocket milling cycle (G78)*

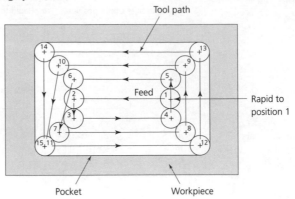

- Repeat movements 2 to 15 inclusive to finish machine bottom and sides of pocket. Note that the number of movements when roughing and finishing will depend upon the diameter of the cutter and the size of the pocket. In actual practice, the paths may overlap by as much as 60 per cent of the cutter diameter.
- At point 15 in this example, rapid up to the clearance plane height and return to position 1.

The programmer has to provide the following positional data: the dimensions for the pocket and the cutter, together with the cutting data.

Circular pockets can be generated using the G77 preparatory code. The ability to produce circular pockets and bores immediately, shows a major advantage of CNC milling machines. With conventional, manually operated machines any circular pocket in a milled surface would have to be produced on a lathe or a boring machine, thus requiring the work to be removed from the milling machine and reset in another machine. With a CNC milling machine all the operations of milling, drilling and tapping, and the production of circular and rectangular pockets, circular and rectangular islands, or pockets and islands of complex profiles can all be machined at one setting.

Here are some other examples of canned cycles used on CNC milling machines:

G80	cancels all 'canned cycles'
G82	drilling cycle with dwell
G83	deep-hole drilling cycle;
G84	tapping cycle
G85	boring cycle
G87/88	deep-hole boring cycle

Canned cycles can also be used with CNC lathes and some typical examples are:

G66/67	contouring cycle
G68/69	rough-turning cycle
G81	turning cycle
G82	facing cycle

Fig. 3.8 *Rough turning cycle (G68)*

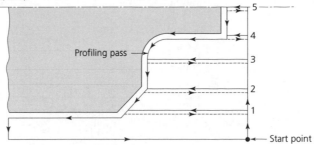

G83   deep-hole drilling cycle
G84/85   straight-threading cycle
G88   auto-grooving cycle

Note how the canned cycles for use on a lathe differ from those for use on a milling machine despite having the same code number. Figure 3.8 shows the sequence of events activated by the G68 preparatory code on a lathe. After the roughing cuts have been made, the profile of the component will be finish turned to size. The sequence of events for the G68 rough-turning cycle is:

• Rapid to the start point.
• Rough out using a sequence of parallel roughing passes. The number of passes will depend upon the depth of cut and the profile.
• A profiling pass leaving a finishing allowance on the component.
• Return in rapid to start point.

3.9.2 *Tool-length offset*

Tool-length offset (TLO) allows tools of various lengths to be used with a common datum, as shown in Fig. 3.9, without having to alter the programme. Setting commences with tool T01, which is 'touched' onto the work surface or a setting block depending upon the Z-axis datum height, and the Z-axis readout is set to zero. The master tool is now T01. Each subsequent tool is then 'touched' onto the Z-axis datum and its Z-axis read-out is noted. Using this information, a Z-axis offset is then applied to each tool in turn to compensate for differences in length compared with T01. The tool-length offsets are recorded in the memory of the machine's computer under the tool number file. Each time a tool is called up by its 'T' code, the correct length offset will be automatically applied. If this parameter changes for a particular tool (the tool is reground), the offset can be reset and no change has to be made to the programme.

The application of tool-length offsets to turning tools is shown in Fig. 3.10. In this case it can be seen that offsets are required in both the X- and the Z-axes relative to a common datum. Usually a number of different tools are located in the lathe turret and each tool will require its own offsets since each tool protrudes by a different distance. The offset for any one tool becomes operative as soon as that tool is called into the programme by its T number.

Fig. 3.9 *Tool length offset: milling*

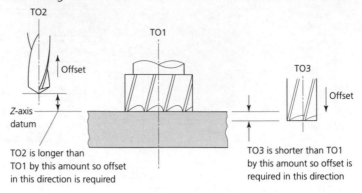

TO2 is longer than
TO1 by this amount so offset
in this direction is required

TO3 is shorter than TO1
by this amount so offset is
required in this direction

Fig. 3.10 *Tool length offset: turning*

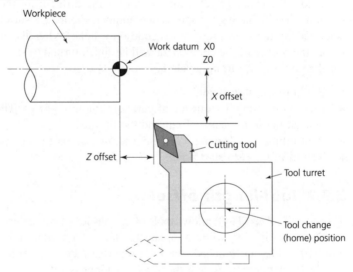

3.9.3 *Cutter diameter and tool-nose radius compensation*

Like tool-length offsets, cutter-diameter compensation (milling) and tool-nose radius compensation (turning) are also facilities provided to aid programming. Not only do these facilities allow tools of different sizes to be interchanged without alteration to the program, they simplify the writing of the program. The tool can be assumed to travel round the profile being machined and allowance for the actual diameter of the cutter is automatically made by the controller. Furthermore the programmer, when programming for turning on the lathe, can assume the tools have a sharp nose. In this instance the controller automatically compensates for the nose radius of the tool.

Fig. 3.11 *Milling cutter diameter compensation: (a) to the left (G41); (b) to the right (G42)*

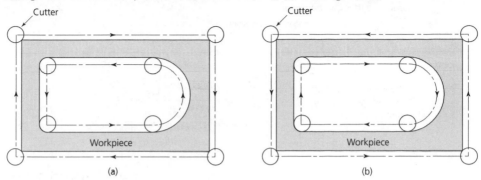

Cutter-diameter compensation for milling machines is controlled by the following preparatory codes:

G41   compensates – cutter to the left of the workpiece [Fig. 3.11(a)]
G42   compensates – cutter to the right of the workpiece [Fig. 3.11(b)]
G40   compensation cancelled

At first sight the 'handing' of the compensation is a little difficult to interpret. Consider Fig. 3.11(a). Start at any point and face in the direction of the cutter travel by following the arrows. It can be seen that the path of the cutter is always to the left of the surface being machined. Similarly, in Fig. 3.11(b) the path of the cutter is always to the right of the surface being machined. The path of the cutter traverse is determined by whether up-cut or down-cut (climb) milling techniques are used.

Whenever the G41 or G42 code is activated, the cutter-diameter compensation is always applied at the next move in the X- and Y-axes, as shown in Fig. 3.12(a), and always cancelled at the next X and Y move after the G40 code is activated, as shown in Fig. 3.12(b). Diameter compensation can never be applied or cancelled while cutting is taking place.

Notice that the preliminary movement of the cutter takes place clear of the work so that compensation is fully effective and the cutter has achieved its required feed rate by the time the cutter is in contact with the workpiece. This allows the feed servomotor to accelerate up to speed and is referred to as *ramping on*. Similarly, at the end of the cut, the cutter feeds clear of the work as the feed servomotor decelerates when the compensation is cancelled. This is referred to as *ramping off*.

Lathe tool-nose radius compensation is used in a similar manner to diameter compensation when milling and the same preparatory codes are used. That is:

G41   tool-nose radius compensation to the left [Fig. 3.13(a)]
G42   tool-nose radius compensation to the right [Fig. 3.13(b)]
G40   tool-nose compensation cancelled

The use of tool-nose radius compensation simplifies programming as it allows the programmed tool-nose path to follow the profile of the component and also allows the programmer to assume a sharp-nosed tool is being used. Like tool-length offsets,

Fig. 3.12 *Cutter compensation: (a) ramping on (G41) initiates cutter compensation; (b) ramping off (G40) cancels cutter compensation*

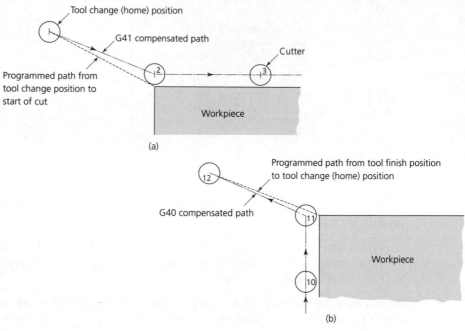

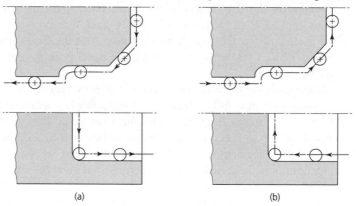

Fig. 3.13 *Lathe tool-nose radius compensation: (a) G41, facing tool always moves to the left of the surface being turned; (b) G42, facing tool always moves to the right of the surface being turned*

diameter compensation and tool-nose compensation are set on the machine itself when the actual tool parameters are known. After the first trial component has been made, the compensation settings may need to be 'tweaked' to allow for cutter deflection and other variables in order to bring the component within its design tolerances.

3.9.4 *Automatic tool changing*

There are several systems of automatic tool changing, but the most popular are indexable tool turrets and chain magazines. The tools are kept in specific positions in the turret or magazine and the appropriate tool is selected and changed as required by the controller in response to program commands. See also Section 3.10.13.

3.9.5 *Writing a simple part program*

Before a part program can be written one further decision has to be made: that is, to select either absolute or incremental dimensioning. Absolute dimensions are the more commonly used as the dimensioning is related to a common datum, as shown in Fig. 3.14(a). Incremental dimensioning is shown in Fig. 3.14(b).

Fig. 3.14 *Dimensioning techniques: (a) absolute; (b) incremental*

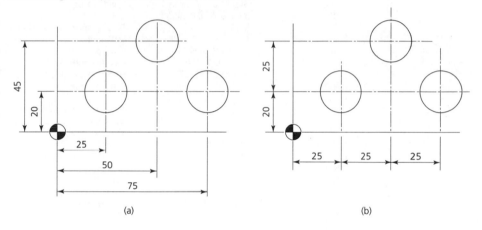

Fig. 3.15 *Component to be milled and drilled on a CNC machine*

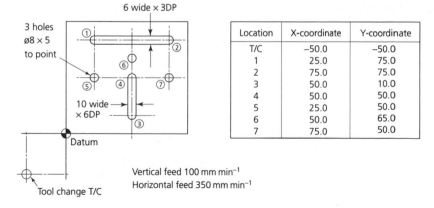

Location	X-coordinate	Y-coordinate
T/C	−50.0	−50.0
1	25.0	75.0
2	75.0	75.0
3	50.0	10.0
4	50.0	50.0
5	25.0	50.0
6	50.0	65.0
7	75.0	50.0

Vertical feed 100 mm min⁻¹
Horizontal feed 350 mm min⁻¹

A simple component to be made on a CNC milling machine is shown in Fig. 3.15. The material is low-carbon steel and the blank to be machined is 10 mm thick. A typical part program for milling and drilling this component is shown in Fig. 3.16.

A simple turned component is shown in Fig. 3.17, and a typical part program for this component is shown in Fig. 3.18.

Fig. 3.16 *Typical part program for machining the component in Fig. 3.15*

Sequence number	Code program	Explanation
%	G00 G71 G75 G90	Default line
N10	X–50.0 Y–50.0 S1000 T1	
N20	M06	TC posn spindle/tool offset (ø6 mm)
N30	X25.0 Y75.0 Z1.0	Rapid posn (1)
N40	G01 Z–3.0 F100	Feed to depth
N50	G01 X75.0 F350	Feed to posn (2)
N60	G00 Z1.0	Rapid tool up
N70	X–50.0 Y–50.0 S800 T2	
	M06	TC posn spindle/tool offset (ø10 mm)
N80	X–50.0 Y10.0 Z1.0	Rapid posn (3)
N90	G01 Z–6.0 F100	Feed to depth
N100	G01 Y50.0 F350	Feed to posn (4)
N110	G00 Z10	Rapid tool up
N120	X–50.0 Y–50.0 S1100 T3	
	M06	TC posn spindle/tool offset (ø8 mm drill)
N130	X25.0 Y50.0 Z1.0	Rapid posn (5)
N140	G81 X25.0 Y50.0 Z–7.0 F100	Drill on restate depth (inc) feed
N150	X50.0 Y65.0	Posn (6)
N160	X75.0 Y50.0	Posn (7) drill
N170	G80	Switch off drill cycle
N180	G00 X–50.0 Y–50.0 M02	TC posn end of prog
E		End of tape

Fig. 3.17 *Component to be turned on a CNC lathe (dimensions in millimetres)*

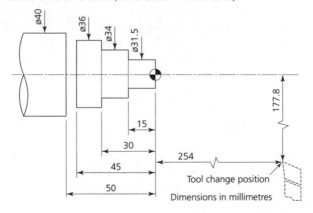

Fig. 3.18 *CNC part program for turning the component in Fig. 3.17 – machine: Hardinge lathe controller GE 1050*

Sequence number	Code program	Explanation
N10	G71	Metric
N20	G95	Feed in mm rev
N30	G97 S1000 M03	Direct spindle. 1000 rev min^{-1} Spindle on CW
N40	G00 M08	Rapid mode. Coolant on
N50	G53 X177.8 Z254 TO	To tool change position
N60	M01	Optional stop
N70	T100	Rotate turret Pos 1
N80	G54 X0 Z2 T101	Move to start with tool 1
N90	G01 Z–0.5 F0.2	Move to depth prior to face end
N100	X31.5	Face end
N110	Z–15	Turn 031.5
N120	X34	Face edge
N130	Z–30	Turn 034
N140	X36	Face edge
N150	Z–50	Turn 036
N160	G53 X177.8 Z254 TO	To tool change
N170	T400	Rotate turret Pos 4
N180	G54 X40 Z–45 T404	To part off position. Tool 4 offset 04
N190	G01 X–1.0 F0.1	Part off
N200	G00 X40	Retract
N210	G53 X177.8 Z254 TO	To tool change
N220	M02	End program

3.10 Computer numerical programming – more advanced part programming

3.10.1 *Further canned cycles*

The G81 drilling cycle was introduced in Section 3.9.1. A 'pecking' infeed was not included in G81 code. *Pecking* is the term used when the drill is fed a short distance into the work and then retracted to clear and break up the swarf before making the next feed increment. The number of pecks will depend upon the depth/diameter ratio of the hole. The deeper the hole, the greater the number of pecks that are required to ensure that the flutes of the drill do not become clogged. This could lead to the drill breaking. Where parameters such as hole depth, retract height, feedrate and *number of pecks* are entered on one line we use the G83 code, for example:

G83 R3 D10 N5 F200

This would produce a hole 10 mm deep, drilled in 5 *pecks*, at a feed rate of 200 mm min^{-1} with a retract height of 3 mm. Other useful examples of canned cycles include:

- tapping screw threads;
- pocket and slot milling;
- area clearance within a boundary;
- turning screw threads.

3.10.2 *Turning example using a subroutine with a loop and a canned cycle*

When a component design results in its program containing fixed sequences of frequently repeated patterns, these sequences can be stored in the control memory as a sub-program, called a *subroutine*, to simplify the task of programming. A subroutine is simply a set of instructions that can be called up and inserted repeatedly into the main body of a program by entering an identifying code.

It is sometimes useful for a subroutine to be repeated a set number of times. The letter P is used to identify the subroutine number. In addition, when using a Fanuc or similar controller, the letter L is used to specify the number of repeats, for example:

N400 M98 P2030 L2

means go to subroutine number 2030 and repeat it twice before returning to the main program. This technique can be illustrated by using a turning program to produce the component shown in Fig. 3.19(a).

The operation is required to turn the 28 mm diameter down to 20 mm diameter in two equal passes of the tool for a length of 50 mm. Because of the simplicity of this example it would normally be performed by a canned cycle on most machines. However, for this example, we will stay with a subroutine. The following program

Fig. 3.19 *Turning component requiring a subroutine with loop: (a) component to be turned; (b) axes for turning. Dimensions in millimetres*

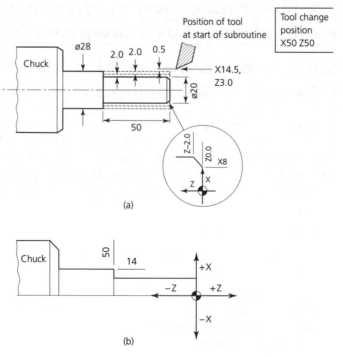

segment assumes that earlier program blocks have set appropriate speed and feeds, set the *tool offsets* (tool-nose radius compensation) and have established the appropriate *absolute positioning mode*, and established the *radius programming mode*. This is because it is essential to tell the controller, at the start, whether program dimensions refer to the *radius* of the workpiece relative to the spindle axis or whether they refer to its *diameter*. As a reminder Fig. 3.19(b) shows the position of the *X* and *Z* dimensions and movements when turning. Remember that movements of the cutting tool *into* the workpiece are always prefaced by a *minus* sign.

Subroutine

:3030	Colon followed by identification number specifies the start of subroutine number 3030
N3040 G91	Incremental programming
N3050 G01 X–2.5	Feedrate move to cutting depth clear of job
N3060 Z–53	Feedrate move to turn diameter
N3070 X0.5	Feedrate move of tool 0.5 mm away from work at end of cut
N3080 G0 Z53	Rapid return with tool 0.5 mm clear of previously turned diameter
N3090 M99	End of subroutine

1. Since the main program will call for the subroutine to be *repeated twice*, the component will be turned to the finished size with 2 mm being taken off the radius (4 mm off the diameter) at each pass of the tool.
2. The tool will be retracted 0.5 mm at the end of each cut so that it does not mark the workpiece as it returns to the start position ready for its next cut. Therefore the tool point is 0.5 mm clear of the work at the start of each pass and this must be added to the required depth of cut.

Let us now see how this subroutine is related to the main program.

Main program

N70 _____	
N80 _____	
N90 X14.5 Z3	Rapid positioning in absolute to start of routine
N100 M98 P3030 L2	Call subroutine 3030 and do twice
N110 G90	On rapid return from subroutine, reset to absolute
N111 G0 X7.0 Z1.0	Rapid to start of chamfer
N112 G1 X11.0 Z–3.0	Machine chamfer at modal feedrate previously set
N120 G0 X50 Z50	Rapid return to tool change position.

Note that the program blocks (lines) are normally numbered in multiples of 10. This allows for additional programming information to be inserted after the program has been written. In the previous example it was decided to include a chamfer on the end of the component to aid a subsequent screw threading operation. Hence the use of the numbers 111 and 112.

3.10.3 *Turning a screw thread using a canned cycle*

Let us suppose that we need to cut a screw thread on the bar we have just turned down to 20 mm diameter. When cutting threads on the end of a bar it is necessary for the tool to start on a chamfer and finish in a groove (a 'landing' groove). This gives the clearance necessary for the cutting tool to begin and end the thread. Remember that the traverse mechanism cannot start and finish instantaneously, therefore there must be room for the tool to accelerate to the correct traverse speed by the start of the thread and to decelerate to a stop at the finish of the cut. While cutting is taking place the rotational speed of the work and the traverse speed of the tool must be synchronised in order that a thread of the correct lead is cut.

The landing groove at the shoulder end of the thread should be equal to or very slightly less than the minor (core) diameter of the thread being cut. This groove is also useful as it allows any associated component, such as a nut, to be tightened up to the shoulder of the bar. Following on from the previous example we now need to change the tool to one suitable for cutting the groove and also change the speed and feed rates to an appropriate value for this operation. The program segment to machine the landing groove is shown in Fig. 3.20.

The procedure for cutting the screw thread is shown in Fig. 3.21. Although the relative rotational movement of the workpiece and the linear movement of the cutting tool *generate* a thread of the correct depth and pitch, a *form tool* is used in order to create the crest radius on the thread and give it the correct form. Note how the cutting tool thread form gets progressively deeper. Figure 3.21 also includes the appropriate program segment for cutting the screw thread based on the Fanuc 0T threading cycle. It follows on from the previous program segment as can be seen from the block numbering, after allowing for a tool change to the threading tool and also changing the speed and feed rates to suit. Note that the pitch of screw thread to be cut governs the feed rate 'F'. Where F1.5 would produce a thread with a 1.5 mm lead (lead = pitch for a single start thread, and lead = pitch × number of starts for multistart threads). Note that the thread gets deeper with each pass of the tool. This is due to the incremental infeed that occurs with each loop of the cycle. Also note how the last cut is termed a *spring cut*.

Fig. 3.20 *Program segment for turning the landing groove*

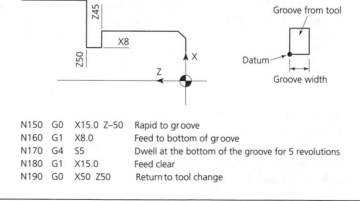

N150	G0	X15.0 Z–50	Rapid to groove
N160	G1	X8.0	Feed to bottom of groove
N170	G4	S5	Dwell at the bottom of the groove for 5 revolutions
N180	G1	X15.0	Feed clear
N190	G0	X50 Z50	Return to tool change

Fig. 3.21 *Program segment for cutting a screw thread*

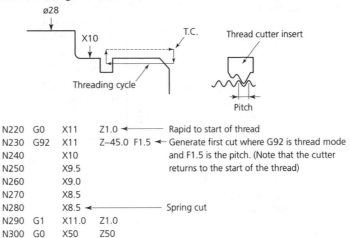

N220	G0	X11	Z1.0 ◄─────── Rapid to start of thread
N230	G92	X11	Z–45.0 F1.5 ◄─ Generate first cut where G92 is thread mode
N240		X10	and F1.5 is the pitch. (Note that the cutter
N250		X9.5	returns to the start of the thread)
N260		X9.0	
N270		X8.5	
N280		X8.5 ◄─────── Spring cut
N290	G1	X11.0	Z1.0
N300	G0	X50	Z50

It is at the same depth as the previous cut and it is there to accommodate any deflections of the cutting tool and the workpiece caused by the cutting forces.

The main advantage gained from the use of subroutines and canned cycles is the time saved in programming, particularly if there is a need for the programmer to write out repetitive blocks of programming that provide for the tool to make a series of identical moves at different stages in the machining.

3.10.4 *Milling example using subroutines*

Figure 3.22(a) shows a fixture plate that consists of 24 holes in a plate 120 mm by 210 mm by 20 mm thick. Each hole is tapped with an M10 screw thread. Figure 3.22(b) shows the program structure for producing the holes. You can see that it has *subroutines* branching from the *main program*. This saves the programmer from having to keep repeat much of the coded information.

3.10.5 *Nested loops*

A nested loop is a repetitive section of program that is contained (nested) within a subroutine. This is often referred to as a *loop within a loop*. For example it is very useful for drilling a matrix of holes, using linear incremental programming, where each hole is drilled relative to the previous position. Returning to our fixture plate with its matrix of holes, we see from the program structure, Fig. 3.22(b), that there is a *nested loop* within the second subroutine.

Finally, to conclude our introduction to subroutines and nested loops, Fig. 3.23 shows the full program for our fixture plate. This has been written for a *Fanuc 0MA* controller. Note that programs written for one make and type of controller will not work correctly on any other make and type of controller. *In the interests of **safety** do **not** attempt to use this program on a machine with any other controller.* This also applies to all other

Fig. 3.22 *Drilling and tapping a matrix of holes in a fixture plate: (a) component to be drilled; (b) program structure*

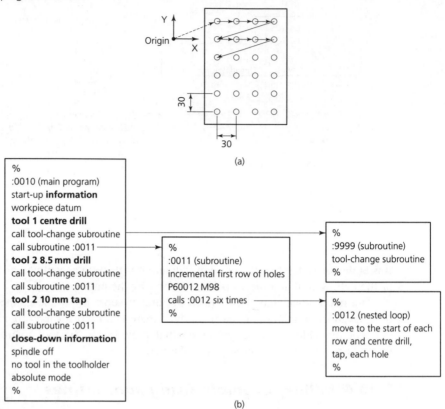

(a)

(b)

programs or segments of programs included in this chapter. They will only work correctly and safely in conjunction with the controller for which they were written.

3.10.6 *Macro language*

Most CNC controllers have a macro programming language. Fanuc 0M controllers have Macro A or Macro B. The macro language allows the user to store variables within registers, for example #100, and use mathematical functions such as sine, cosine, square root, etc. This language is like other conventional computer languages such as BASIC, and can be used for more sophisticated applications such as:

- positioning holes on a pitch circle (PCD);
- elliptical path machining;
- in-cycle gauging;
- setting up tools and workpieces using probes.

For example, let us suppose a company has a frequent need to machine bolt holes on a pitch circle with equal hole spacing but that the radius of the pitch circle and the

Fig. 3.23 *Full program for fixture plate*

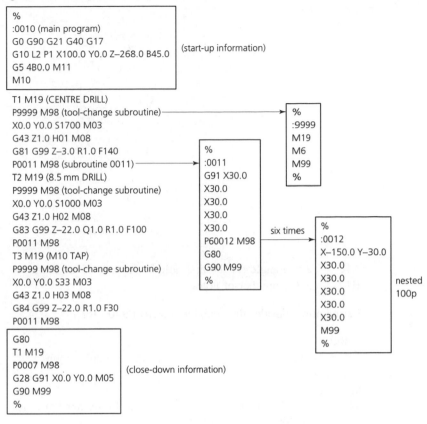

```
%
:0010 (main program)
G0 G90 G21 G40 G17
G10 L2 P1 X100.0 Y0.0 Z-268.0 B45.0          (start-up information)
G5 4B0.0 M11
M10
```

```
T1 M19 (CENTRE DRILL)
P9999 M98 (tool-change subroutine)                              %
X0.0 Y0.0 S1700 M03                                             :9999
G43 Z1.0 H01 M08                                               M19
G81 G99 Z-3.0 R1.0 F140               %                         M6
P0011 M98 (subroutine 0011)          :0011                      M99
T2 M19 (8.5 mm DRILL)                G91 X30.0                  %
P9999 M98 (tool-change subroutine)   X30.0
X0.0 Y0.0 S1000 M03                  X30.0
G43 Z1.0 H02 M08                     X30.0
G83 G99 Z-22.0 Q1.0 R1.0 F100        X30.0          six times    %
P0011 M98                            X30.0                      :0012
T3 M19 (M10 TAP)                     P60012 M98                 X-150.0 Y-30.0
P9999 M98 (tool-change subroutine)   G80                        X30.0
X0.0 Y0.0 S33 M03                    G90 M99                    X30.0          nested
G43 Z1.0 H03 M08                     %                          X30.0          100p
G84 G99 Z-22.0 R1.0 F30                                         X30.0
P0011 M98                                                       X30.0
                                                               M99
G80                                                            %
T1 M19
P0007 M98                     (close-down information)
G28 G91 X0.0 Y0.0 M05
G90 M99
%
```

number of holes varies from job to job. If the machine controller does not have a pitch circle canned cycle, then the company may find that it is worth writing a suitable *macro* and storing it permanently in the controller memory.

Figure 3.24 shows a typical bolt-hole pitch circle with its parameters. This macro would enable calculations to be performed for the coordinates of each hole on the pitch circle for any particular values of A, R and H needed for a given job, within the machining program. The programmer would merely need to enter a single line statement containing the required values of A, R, and H, to cause the coordinate values to be calculated. A suitable canned cycle would then be used to perform the actual machine movements. The macro call statement takes the following form for machining 12 holes on a pitch circle of 30 mm radius:

G65 P9400 R30 A10 H12

where:

G65	code calls for the macro
P9400	is the macro number (same format as for the subroutine)
R30	is the radius of the pitch circle (mm)

Fig. 3.24 *Holes on a pitch circle: (X0, Y0) = coordinate value of bolt-hole circle centre; R = radius of bolt-hole circle; A = start angle*

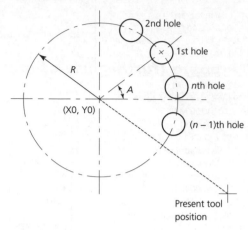

A10   is the angle for the first hole in degrees of arc from the origin

H12   is the number of holes

Let us now consider the complete macro P9400 for Fig. 3.24.

```
:9400
#30 = #101;                          Store reference point
#31 = #102;
#32 = 1;
While [#32 LE ABS(#11)] Do 1;        Repeat by number of holes
#33 = #1 + 360 * [#32–1] / #11
#101 = #30 + #18 * cos[#33];         Hole position
#102 = #31 + #18 * sin[#33];
X#101 Y#102;
#100 = #100 + 1;                     Increase hole count by 1
#32 = #32 + 1;
End 1
#101 = #30                           Return to reference point
#102 = #31
M99                                  End of macro
```

Note that numbers with # in front are variables in the normal computing sense, where variables have the following meanings:

    #100   Hole number counter

    #101   X-coordinate reference point

    #102   Y-coordinate reference point

    #18   Radius *R*

    #1   Start angle *A*

    #11   Number of holes *H*

#30   Storage of X-coordinate reference point
#31   Storage of Y-coordinate reference point
#32   Counter for the nth hole
#33   Angle of the nth hole

3.10.7 *Probing (further use of macros)*

The touch-trigger probes used on CNC machines are different from and more robust in construction than those used for measurement since they are permanently mounted in the machine carousel or magazine. The benefits of probing can be listed as follows:

- Reduced machine down time.
- Reduced operator attendance.
- Approved product quality.
- Increased productivity.
- Increased profitability.

The touch-trigger probe records the coordinates of the position of the probe stylus at the moment that it touches a surface. These coordinate values are saved within a register and compared to the desired value stored within another register. The difference between these two values is the *error value* that may be used to update and correct a workpiece datum or a tool-radius or length-offset value. The stylus of the probe is connected to a trifilar suspension system that is supported on three rollers. It is the change in electrical resistance as the rollers begin to separate, and not the actual breaking of contact between the rollers, that triggers the probe. The signal is transmitted from the probe by an optical, an inductive or a radio transmission system. The signals from the probes are picked up by means of appropriate receivers. Like any other tool, the probe is called up by the program as and when it is required. It is activated by being spun in one direction to switch it on (P9001) and then, after use, it is spun in the opposite direction to switch it off (P9002) before it returns to its storage position in the carousel or tool magazine.

Let us now consider an example of how a macro segment is used to probe a Z datum, as shown in Fig. 3.25.

:0001   Program number one

Call up the probe

P9001   Spin the probe to turn it on
P9005   Clear the macro registers

Fig. 3.25 *Use of probe to establish height of a surface above the Z datum*

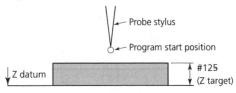

Move the probe to the start position

G65 H1 P#100 Q21	Start of macro: Assign metric units
G65 H1 P#126 Q10000	Target at Z10.00 (note: Q dimensions in μm)
G65 H1 P#119 Q12	Update the workpiece datum P2 from the reference datum P1
P9018 M98	Call up the Z measure macro

Move the probe away

P9002	Spin the probe to turn it off

This macro is used to determine automatically the *actual* datum surface (Z = 0) of the workpiece. This is necessary when components having different thicknesses have to be machined using a common programme. No matter what the thickness of the workpiece, once the probe has touched its upper surface, that surface then becomes the new datum.

3.10.8 *Scaling*

The scaling feature allows the X- and Y-coordinates in milling and drilling, and the X- and Z-coordinates in turning, to be increased or decreased by a scaling factor from their stated values in the program. For example, Fig. 3.26 shows a component requiring two slots to be machined with their widths equal to the cutter diameter. To simplify the programming, a subroutine could be written to machine the inner slot and this could be called twice. Once *without* the scaling facility active to machine the inner slot, and a second time *with* the scaling facility active with a scaling factor of 1.5 (150/100) to machine the outer slot. Alternatively, the subroutine could be written for the outer slot and this could be called a second time with a scaling factor of 2/3 active to machine the inner slot.

Fig. 3.26 *Scaling*

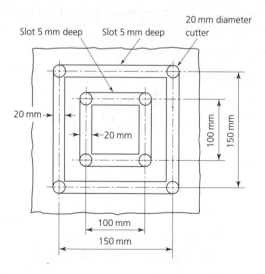

Fig. 3.27 *Mirror imaging*

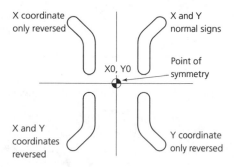

X coordinate only reversed

X and Y normal signs

Point of symmetry

X0, Y0

X and Y coordinates reversed

Y coordinate only reversed

3.10.9 *Mirror imaging*

Mirror imaging is a useful feature, available with some control systems, that allows either a whole program or a subroutine to have the signs of its coordinate data selectively reversed. The facility is applicable to milling and drilling operations and X-coordinates or Y-coordinates or both may be reversed.

Let us consider Fig. 3.27. This shows a cavity that has to be machined into a die block at four symmetrical positions. If the program datum (point of symmetry) is defined in the position shown (X0,Y0) the subroutine to produce the cavity would normally result in the top right-hand quadrant being machined first. If the subroutine is then called with the X reversal facility active, then the top left-hand quadrant would be machined. If the subroutine is called with the X and Y reversal facilities active, then the bottom left-hand quadrant would be machined. Finally, if only the Y reversal facility is active, then the bottom right-hand quadrant would be machined. Clearly, a significant amount of programming time can be saved using the mirror image facility and there is also less opportunity for the introduction of programming errors.

The International Standards Organisation (ISO) code for mirror imaging is G28 but control systems vary greatly in their adherence to standards, and other methods of activating the feature may be encountered. Assuming that the control system does use G28, the following program lines illustrate how mirror imaging may be called up. For example:

N110 G28 X	This reverses the sign of the X-coordinates only for the subsequent subroutine
N120 to N180	The subsequent subroutine
N190 G28 Y	This reverses the sign of the Y-coordinates only for the subsequent subroutine
N200 to N280	The subsequent subroutine
N290 G28 XY	This reverses the signs of both the X- and the Y-coordinates for the subsequent subroutine.

3.10.10 *Rotation*

Rotation, which is available with some control systems, is a useful feature that allows the whole coordinate system to be rotated by a stated angle. This facility is applied

Fig. 3.28 *Rotation*

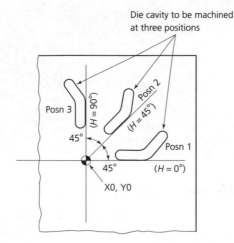

to milling and drilling operations and is particularly useful when a machine feature is repeated at various angular positions around a common centre as shown in Fig. 3.28.

The most efficient way to prepare a program to machine this component would be to write a subroutine that carries out the machining of the cavity, and then to use the *rotation feature* to rotate the coordinate axis system each time the subroutine is called. The ISO word address code for rotation is G73, with the H address used to store the rotation angle. However, these standards are not universally applied and other methods of activation may be met. Assuming the ISO system to be in use, the general form of the program to machine the component shown in Fig. 3.28 would be:

N200 call subroutine to machine cavity	Causes cavity to be machined at position 1
N210 G73 H45	Rotates coordinate system by 45° from angle zero
N220 call subroutine to machine cavity	Causes cavity to be machined at position 2
N230 G73 H90	Rotates coordinate system by 90° from angle zero
N240 call subroutine to machine cavity	Causes cavity to be machined at position 3
N250 G73 H0	Sets coordinate system back to normal (0°)

3.10.11 *Zero shift*

The zero shift facility is commonly available on most control systems. It allows the program datum to be changed within the program. The component shown in Fig. 3.29 will be used to explain this feature. A program has to be written to machine the large hole A and then the smaller holes B that are arranged in a pattern about a centre point.

Fig. 3.29 *Zero shift (dimensions in millimetres)*

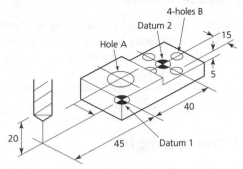

The most convenient datum for hole A is datum 1 and a convenient tool-change position is X−45, Y0, and Z20 with reference to datum 1. The first line of the program would therefore define the start position of the tool (the tool-change position) using a G92 code and the machine operator would position the tool at this point at the start of the program. For example:

N100 G92 X−45 Y0 Z20	Define tool change position relative to datum 1
————	
————	Program lines for machining hole A
————	
N1200 G00 X−45 Y0 Z20 M06	Return to tool change position and change tool

In order to minimise the calculations for the pattern of holes B the adoption of datum 2 would be most convenient. Therefore the program would continue with a G92 code to tell the control system where the existing tool-change position is relative to datum 2. This continuation of the program would read:

N1250 G92 X−85 Y−15 Z20	Define tool-change position relative to datum 2
———	
———	Program lines to machine the four smaller holes
———	
N3000 G00 X−85 Y−15 Z20 M02	Return to tool-change position at end of program

We have already drawn attention to the fact that a programme for one controller will not work on another. For example, when discussing screw-thread cutting we used a G92 code for the appropriate canned cycle. This was because the program segment was written to work with a Fanuc 0T controller. For this milling example we have assumed the use of a controller operating with ISO codes. Therefore we have used a G92 code for position preset.

3.10.12 *Block delete*

The block delete feature is used to omit parts of programs that may not be required. For example two components may be identical except for a single hole. Sometimes the hole will be required and sometimes it will not be required. Rather than write two programs, a single program incorporating the hole can be used. When the hole is not required, the operator can press the 'block skip' button and the block of code with '/' in front of it will be skipped over, and the hole will be omitted.

3.10.13 *Tool changing*

Tool changing can be performed manually or automatically. Since automatic tool-changing facilities add considerably to the cost of a machine, manual tool changing is still widely used for small quantity batch-work. When manual tool changing is used it is essential to minimise the changeover time. This can be achieved by using quick-change tool holders and pre-set tooling. The tools are kept in a 'crib' placed conveniently beside the machine. The 'crib' is a stand in which the tools are not only stored when not in

Fig. 3.30 *Automatic tool changing: (a) chain magazine; (b) rotary turret magazine*

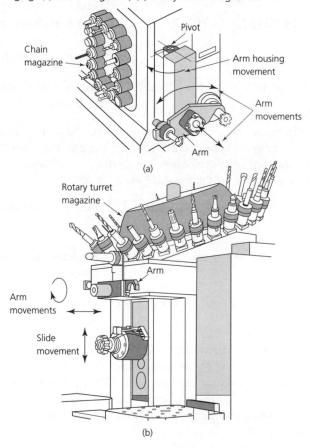

use, but are located in the order in which they are to be used, and in such a position that they can be easily grasped and removed from their location.

Automatic tool-changing systems are classified according to the way in which the tools are stored.

Indexable turrets can be programmed to rotate (index) so as to present the tools mounted in the turret in the order in which they are required. Indexable turrets are widely used on turning centres and also on some milling and drilling machines. In the latter case, the drive to the cutting tool is also transmitted through the turret. Such a system lacks the rigidity of a conventional machine head and spindle assembly and is usually only used on comparatively-light-duty machines.

Tool magazines are indexable storage facilities and are used only on machining centres. Two systems are shown in Fig. 3.30. The tool magazine is indexed to the tool-change position and an arm removes the current tool from the machine spindle and inserts it into the empty socket in the magazine. The magazine then indexes so that the next tool to be used is presented to the tool-change position. The arm then extracts the tool-holder and tool from the magazine and inserts it into the machine spindle ready for use.

3.11 Computer aided design and manufacture

The previous sections have shown how CNC machine tools can be programmed to machine a variety of shapes. However, as mentioned in Chapter 1, design decisions must be made with manufacturing implications in mind, i.e. *concurrent engineering philosophy*. At a basic level using the same geometry as that created by the designer in a computer aided design (CAD) package rather than recreating it in a computer aided manufacturing (CAM) package is important in terms of both accuracy and time saving. The different levels of complexity can range from simple two-dimensional profiles for flame cutters to the complex multi-axis machining of dies from three-dimensional surface models. The advent of rapid prototyping technologies is a very important development where concept models can be grown directly from the design geometry or machined directly on a desk-top milling machine. Figure 3.31 shows the integration of each area.

3.11.1 *CAD systems*

Industry standard packages, such as AutoCad, can export DXF (Drawing Exchange File) geometry files in a format that can be imported by many CAM packages. Other file

Fig. 3.31 *Relationship of CAD/CAM functions*

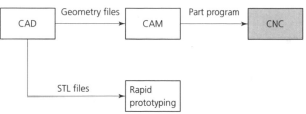

formats used for exporting by CAD packages include IGES (Initial Graphical Exchange System), which is also a neutral file format (can be read by many systems without translation). The important point to note when creating geometry in a CAD package is to ensure that the *world origin* is at the bottom left-hand corner of the component. This ensures that the world origin is aligned with the workpiece datum of the component that is going to be simulated in the CAM package.

3.11.2 *CAM systems*

Computer aided manufacturing systems include PEPS (Production Engineering Productivity System) that can import DXF files. These, in turn, can be defined as part boundaries (outer profiles and inner pockets or slots) called K curves. These boundaries are usually defined in a clockwise direction for milling as they are used to indicate the direction of the milling cutter. The language used in PEPS is very powerful, with one-line commands defining pockets or profiles. Once the program has been proved, i.e. the simulation of the cutter paths is correct, then the simulation can be *post-processed* into any CNC language, i.e. Heidenhain, Fanuc, Allen Bradley, Phillips.

3.11.3 *PEPS programming language*

Let us consider the example shown in Fig. 3.32. This has four four curves (K1, K2, K3 and K4) and one group of holes (G1). The geometry was created by importing a DXF file for the component drawn in AutoCad. In the PEPS geometry menu the DXF lines

Fig. 3.32 *PEPS example drawn in AutoCAD*

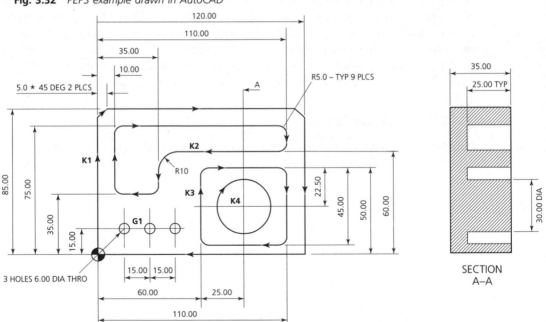

can be converted to a K curve simply by selecting, in a clockwise direction around each boundary, 'K curve command' for each line in turn. Finally E is selected to end and close the boundary. The group of holes is defined in a similar way. 'Group' is selected from the geometry menu and then each individual point defined within the group is selected. The final geometry is defined thus:

K1 is the outer profile
K2 is the long inner pocket
K3 and K4 are the square pocket with a circular island
G1 is a group of three holes to be drilled

Let us now consider the machining commands to produce K1. These commands are defined as follows:

TOOL 1 D20	is a 20 mm diameter cutter
FRO X–37.5 Y 28.5	is the tool-change position
SPI 1250	is the spindle speed in revolutions per min
FED V150 H200	is the vertical and horizontal feedrates in mm per min
CLE 3	is a clearance plane that the cutter rapids up and down to
RAP	is rapid feedrate
RET	retracts the tool to the clearance plane
DES –15	descends at feedrate 15 mm below the part surface
OFF L0	offsets the cutter to the left of the profile by the cutter radius
PRO TK1	profiles K1 in the direction it was defined (tangential), in this case a clockwise direction
RET	retracts to the clearance plane
RAP	rapid feedrate
SPI 0	spindle off
GOH	means go home to the tool-change position

Note: The start up and close down information is defined in the same safe manner as in conventional CNC programming. This simulation is shown in Fig. 3.33.

Fig. 3.33 *Profiling cutter path*

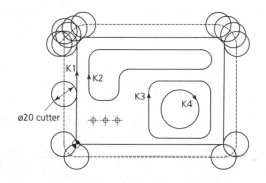

3.11.4 *Simulation of automated systems*

Simulation of tool paths is useful before machining a component on a very expensive CNC machine tool, for example:

- the operator can be certain that the tool is not going to make any unpredicted moves;
- programs can be prepared off-line to avoid wasting valuable machine time;
- safe clamping positions can be determined;
- programmers only need to learn one language irrespective of the number of different machine controllers that may be on site;
- programming using simulation is simpler and quicker for complex components and does not require calculations that might lead to human error, which could result in damage to expensive plant or injury to the operator.

ASSIGNMENTS

1. Despite the fact that CNC machine tools are more complex and costly than conventional machine tools, they are being used increasingly in manufacturing industry. Discuss reasons why this is so.

2. With the aid of sketches explain the difference between open-loop and closed-loop control systems, and list the advantages and limitations of each system.

3. Explain what is meant by the following terms used in programming:
    - (i) character;
    - (ii) management word;
    - (iii) dimensional word;
    - (iv) block;
    - (v) preparatory code;
    - (vi) miscellaneous command;
    - (vii) modal command.

4. (a) State what is meant by the term *canned cycle*.
    (b) State the advantage of using canned cycles.
    (c) With the aid of sketches show the sequence of a typical canned cycle.

5. (a) Describe the uses and benefits of subroutines in CNC part programming.
    (b) Differentiate between macros and subroutines in CNC programming.
    (c) Describe a typical machining situation where a macro-programming facility would be useful and explain in general terms how the macro would work.

6. List the main stages in the preparation of a CNC part-program using the PEPS computer aided part programming system and briefly explain what happens at each stage.

7. With the aid of simple sketches, explain the following CNC part programming features:
    - (a) zero shift;
    - (b) scaling;
    - (c) rotation;
    - (d) mirror imaging.

Fig. 3.34 *All dimensions in millimetres; surfaces and edges of blank previously machined to size (90 × 55 × 20); material is 0.5% plain carbon steel*

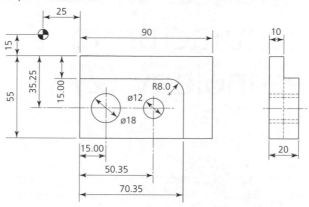

Fig. 3.35 *All dimensions in millimetres; blank ø55 × 72 one end already faced; material is Duralumin*

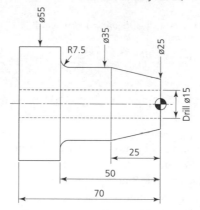

8. (a) Prepare an operation planning sheet listing the tooling, spindle speeds and feedrates for machining the component shown in Fig. 3.34.

   (b) Use this data to prepare a program for a part for machining the component from the solid on a vertical spindle CNC milling machine. Specify the controller and the machine for which the program has been written.

9. (a) Prepare an operation planning sheet listing the tooling, spindle speeds and feedrates for turning the component shown in Fig. 3.35.

   (b) Use this data to prepare a program for a part for machining the component from a solid blank on a CNC lathe. Specify the controller and the machine for which the programme has been written.

4 Advances in manufacturing technology (2)

When you have read this chapter you should be able to:

- describe how the application of CNC has influenced surface-grinding processes;
- describe how the application of CNC has influenced cylindrical-grinding processes;
- describe how the application of CNC has influenced centreless-grinding processes;
- appreciate how CNC has influenced sheet metal punching operations;
- describe how CNC has influenced metal forming operations such as hydro-forming and tube bending;
- appreciate the application of CNC to non-engineering manufacturing processes.

The computer control of turning centres and machining centres was considered in some depth in the previous chapter. Partly because it was with turning and milling processes that e-manufacturing technology started and partly because once the principles of part programming are thoroughly understood they can be applied to many additional engineering and non-engineering processes. So let us now look at some of these additional processes.

4.1 Precision grinding

The principles of conventional surface and cylindrical grinding and the use of abrasive wheels for metal removal were considered in *Manufacturing Technology*, Volume 1, and centreless grinding was introduced in Volume 2. Just as computer numerical control has taken over from the manual and mechanical control of conventional lathes and milling machines for batch and volume production, so it has taken over many aspects of precision grinding.

4.1.1 *Surface grinding*

Microprocessor control is incorporated in precision grinding machines at various levels of sophistication. For example the original Jones and Shipman model 540, as shown in

Fig. 4.1 *Jones and Shipman 540 surface grinding machine*

Fig. 4.1, has been the backbone of toolroom surface grinding ever since World War II. It has a hydraulic system to traverse the worktable longitudinally and a mechanical cross-traverse mechanism. In-feed of the grinding wheel is under manual control. The stroke of the table is set by adjustable table dogs that engage a hydraulic valve which reverses the direction of table travel at the end of each stroke. There is also a traverse speed control valve which can be adjusted by the operator. With the advent of electronic microprocessor-control devices, this machine is also offered as the 540E and a more sophisticated version the 540X.

The 540E microprocessor-controlled surface grinder offers the accuracy, precision and reliability of the original model 540 but with greater ease of setting and operation. A touch-sensitive control panel with light emitting diode (LED) indicator lights allows the operator to select cycle, cross-feed and table functions, individually or in combination. The cross feed can operate intermittently or continuously and has reversing or single pass and trip facilities. The cross feed is electrically operated and controlled, while the table traverse is hydraulically actuated and electronically controlled with proximity-switch reverse dogs.

The model 540X represents the next stage of sophistication in microprocessor control. The comprehensive control panel provides a choice of three cross-feed modes, automatic continuous, automatic intermittent or manual. Electronic saddle lock and digital display of the saddle position and fine feed rate is also provided, as is auto-saddle reversal. A digital display of the wheel head position and fine feed rate is provided together with electronic two-rate feed with automatic retraction after 'spark out'. In addition there are four standard menu cycles. A standard feature of the 'X' series of machines is the ability to select and display menu cycles. In addition to showing position and feed value, the cross-feed display also allows selection of the series of menu cycles that are as follows.

Menu 1
Menu 1 provides for *run out*. This causes the saddle to run forward for loading when area grinding, if the cycle is completed with the saddle at the rear reversal point.

Menu 2
Menu 2 provides for cylindrical grinding and is used with a cylindrical-grinding attachment. It converts the wheelhead display to show increments as diametral values (i.e. × 2).

Menu 3
On selection of menu 3 the wheelhead increments are multiplied by ten times their previously set value to allow for coarse feed increments.

Menu 4
Menu 4 provides for auto-retract of the wheelhead. The wheelhead retracts into its reset position on completion of the grinding cycle and stroke count, leaving the machine ready for the start of the next grinding cycle.

Finally we come to full CNC control for production surface grinding. A typical example is the Jones and Shipman model 844X – the *TechMaster*. The comprehensive control panel is shown in Fig. 4.2. It can be seen that the control panel is divided into a number of zones, with the controls conveniently grouped and positioned in each zone according to their function.

The grinding modes available under CNC control on this machine are shown in Fig. 4.3 and can be described as follows.

Intermittent cross-feed mode
The cross-feed increment is applied at each *table reversal*, and the down-feed increment is applied at each *cross-feed reversal*. Pre-selected coarse-feed increments, fine-feed increments, spark-out and automatic wheel retraction are all applied under CNC control.

Continuous cross-feed mode
The cross feed moves the table with a continuous feed rate, and the down-feed increment is applied at each *cross-feed reversal*. Pre-selected coarse increments, fine increments, spark-out and automatic wheel retraction are all applied under CNC control.

Plunge and area grind mode
Down-feed increments are applied at each table reversal, with pre-selected coarse increments, fine increments, spark out and wheelhead retraction also applied. The

Fig. 4.2 *The Techmaster CNC control panel*

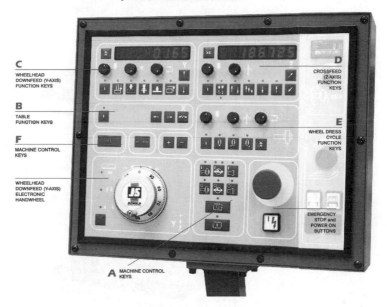

saddle then steps over by a pre-selected amount and the process is repeated. After sufficient 'step-over cycles' have been completed to rough out the surface, the continuous cross-feed mode, as described above, is automatically engaged and the surface area is ground to a fine finish.

Lines and multislot grind modes

Down-feed increments are applied at each table reversal until the required depth of slot is reached. The saddle is locked automatically while this is taking place. No cross feed is applied. When the wheel has reached the pre-selected depth, the wheelhead retracts and the saddle is unlocked. If multiple slots are being ground, the saddle steps over to the position of the next slot and the processes cycle is repeated until all the slots have been ground. The wheelhead finally retracts and the table moves forward to the load/unload position.

4.1.2 *High-volume production surface grinding*

The Jones and Shipman *Dominator* series of production surface-grinding machines has a different geometry to the conventional design of surface-grinding machines. The absence of the reciprocating table and wheelhead column results in structural and dynamic rigidity and high performance. The machine comprises a very substantial cuboid machine bed with the table mounted on the Y-axis supported from the front face on widely spaced

Fig. 4.3 *CNC surface grinding modes*

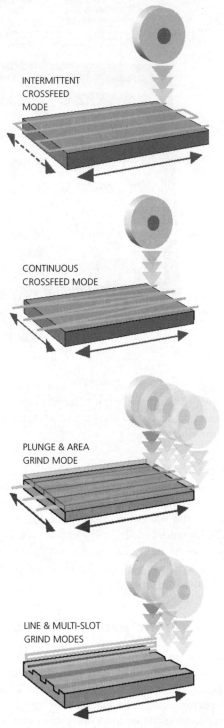

INTERMITTENT
CROSSFEED
MODE

CONTINUOUS
CROSSFEED MODE

PLUNGE & AREA
GRIND MODE

LINE & MULTI-SLOT
GRIND MODES

Fig. 4.4 *Production surface grinding: (a) the Dominator high-volume surface grinding machine; (b) the geometry of the Dominator machine*

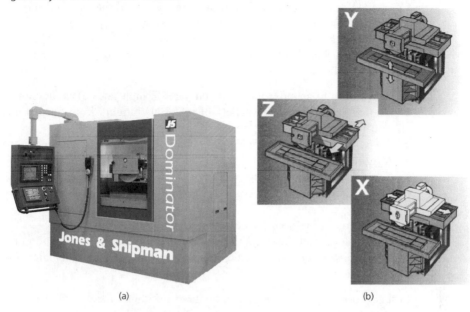

(a)   (b)

preloaded slides as shown in Fig. 4.4. The wheelhead carrier assembly is supported by a saddle mounted on top of the bed that provides the wheelhead 'Z' movement, the wheelhead carrier itself executes the main 'X' movement in either *creep-feed* (see Section 4.1.3) or conventional reciprocating modes. All movements are by recirculating ball screws actuated by GE Fanuc digital servo drives and motors with linear scale feedback.

The level of CNC control in this machine is taken even further. For example table-mounted CNC wheel form-dressing attachments are available for grinding complex profiles. Alternatively, if the production application demands it, table-mounted diamond roll or wheelhead mounted 'overtop' continuous-dress diamond roll units can also be used. Rotary indexing units for inline, vertical or cross-table mounting, with tailstock and wheel conditioning options if required. Punch grinding and automatic component fixturing can all be accommodated on this machine.

The machine is fitted with GE Fanuc CNC intelligent hardware, and a Windows® user interface provides both the machine control and a full suite of wheel dressing and component grinding programs; graphical, colour images prompt the operator and spreadsheet style programming ensures quick and simple input. The option of ISO programming is also available if required.

A rapid-transfer shuttle option allows for in-cycle loading and unloading using an EROWA pallet system for accuracy and reliability. The rapid-transfer option can be configured to accept robotic auto-loading and unloading. The machine is totally enclosed for safe operation, noise reduction and zero coolant pollution.

4.1.3 *CNC creep-feed grinding*

Creep-feed grinding is a CNC controlled process with the standard removal capability of a milling process, combined with the precision, quality and surface integrity associated with the grinding process. This process is widely used for machining the 'fir-tree' roots of gas turbine blades. The major advantages of creep-feed grinding may be summarised as follows:

- The ability to manufacture precision parts at high rates of production.
- The ability to machine a variety of materials ranging from ferrous metals through non-ferrous metals to ceramics.
- Better use of production time by eliminating conventional machining processes prior to grinding.
- Lower consumable tooling costs.
- Lower capital equipment costs by eliminating the need for conventional machining prior to grinding.
- Reduction of the need for de-burring operations.
- By finish grinding from previously heat-treated blanks, elimination of straightening operations to remove distortion caused by conventional-machining stresses during heat treatment.

In this process a free-cutting open-textured grinding wheel is used to reduce the heat generated. Also, a copious supply of coolant is flooded over the cutting zone. Grinding wheel forming and dressing is carried on continuously under CNC control during the grinding cycle. This applies to both conventional and contour grinding. Instead of the rapid table travel and small in-feed increments of conventional surface grinding, this process cuts almost to the full depth in one pass but with a very low table-feedrate more akin to the milling process. Hence the term 'creep feed'.

This machine and its operator communicate by an easy to read liquid crystal display (LCD) colour screen and clearly described keys and controls. Using a combination of graphic and text prompts, the operator is taken through simple, digitising style set-up routines to create wheel forming/dressing and grinding part programs. Suggested parameter values are offered, which can be selected or amended as required.

Each part program consists of a series of custom macros. When a part program needs amending the operator can edit the displayed macro either by using the trim routine or by using the software editing facility, both of which are built into the control menu. Since the machine uses an ISO type format, part programs can be constructed by this method instead of the control menu. Whatever method is used to construct simple or complex machine cycles, the control system is designed to be user friendly.

4.1.4 *Cylindrical grinding*

The same levels of electronic control equally apply to the Jones and Shipman range of cylindrical-grinding machines. These machines range from entry level manually controlled machines, through various stages of electronic microprocessor control in the E and X models, to full CNC control in the *Ultramat*, which is shown in Fig. 4.5 and is suitable for batch and volume production.

Fig. 4.5 *The Jones and Shipman Ultramat cylindrical grinding machine*

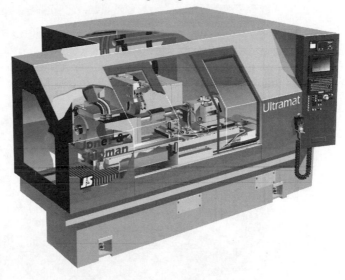

This machine combines independent live and dead spindles in one unit and allows the bearings for the live spindle to be positioned for maximum stiffness. To assist linear positioning an air-lift is provided. A variable speed a.c. drive to the spindles with an encoder ensures fast and accurate spindle orientation. For taper correction in the live grinding mode a fine angle adjustment facility is provided, with its position displayed on the screen. Thermal stability in the work area is ensured by porting the coolant through the structure of the workhead, wheelhead and tailstock. A high-accuracy taper correction tailstock is also provided.

The control pendant contains:

- an alpha-numeric keyboard;
- an industrial fixed mouse;
- buttons for main machine control and mode selection;
- wheel balancing and gauging display units.

The remote handset contains:

- an HPG (hand pulse generator) handwheel;
- manual axes selection;
- feedrate override;
- emergency stop.

Windows® set-up pages allow the operator to quickly digitise the diamond wheel dresser and wheel positions and, using a minimum of mouse or softkey/keyboard inputs, produce a finished program. Spreadsheet style programming ensures quick and simple input. Graphical images prompt the operator. In-process gauging systems are available, including: diameter, face flag and probe. A probe can be seen controlling the part diameter in the external grinding operation shown in Fig. 4.6. The machine

can be configured for autoloading and tailored to suit customers' specific requirements and preferred makes of load/unload equipment. The machine is totally enclosed for safety, noise control and zero coolant pollution. Automatic door opening and closing is also available. To increase production rates still further, 'gap-elimination' is provided. Sonic sensors that are built into the machine listen to the changing note of the coolant trapped between the wheel and the workpiece so that the rapid in-feed is stopped when the wheel closes the gap but before it commences to cut. The incremental in-feed is then applied by the CNC controller and the programmed grinding cycle takes place.

4.1.5 *Superabrasive machining*

Superabrasive machining is neither true milling nor true grinding but combines the attributes of both processes in that rapid material removal is possible while maintaining the finish and accuracy of a ground surface. The cutting wheels have a pre-formed steel core and are electroplated with cubic boron nitride (CBN). This is a synthetic abrasive that is as hard as diamond. At high wheel-speeds and relatively high feedrates, the CBN wheels provide surface finishes equal to grinding. For example the components shown in Fig. 4.7 can be machined from the solid in a heat-treated (hardened and

Fig. 4.7 *Superabrasive (CBN) grinding*

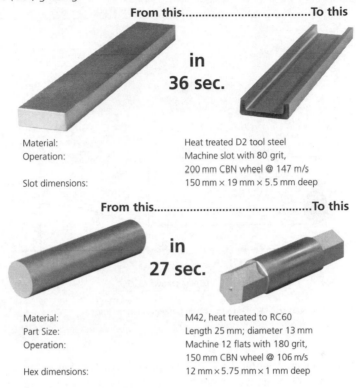

From this..To this

in
36 sec.

Material:
Operation:

Slot dimensions:

Heat treated D2 tool steel
Machine slot with 80 grit,
200 mm CBN wheel @ 147 m/s
150 mm × 19 mm × 5.5 mm deep

From this...To this

in
27 sec.

Material:
Part Size:
Operation:

Hex dimensions:

M42, heat treated to RC60
Length 25 mm; diameter 13 mm
Machine 12 flats with 180 grit,
150 mm CBN wheel @ 106 m/s
12 mm × 5.75 mm × 1 mm deep

tempered) alloy steel. To take full advantage of this machining process, CNC machines that are both powerful and robust are required.

Superabrasive CBN cutting tools can be used to cut both soft and hard materials, from free-machining non-ferrous metals to super alloys such as Inconel and Monel metals. In addition advanced materials such as ceramics and composites can be machined equally well using the superabrasive machining system. Tests have proven that the forces induced onto the workpiece are much less than with creepfeed grinding using conventional abrasive wheels and milling. As a result, there is virtually no deformation of the workpiece. The superabrasive wheels are economical in use since the pre-formed steel core is of relatively low cost, while significant stock removal can be achieved before the wheel needs to be stripped and replated with CBN. Typical wheel life can be as high as 5735 cm^3 of stock removal.

4.2 Centreless grinding (conventional)

Before we can consider the application of CNC to the centreless-grinding process, we need to consider the principles of the process itself. The cost of conventional cylindrical-grinding processes where the work has to be mounted in chucks or between centres, as

Fig. 4.8 *Centreless ground flexure spring*

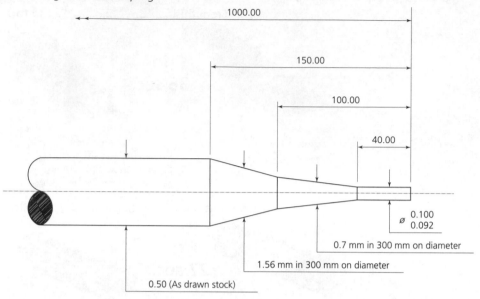

Note: Diameters scale 50 : 1; Lengths NTS; Dimensions in millimetres.

described in Section 4.1.4, can be prohibitive for many applications. Furthermore for some components it may not be a suitable process because the workpiece is too slender to withstand the grinding forces. A component such as that shown in Fig. 4.8 can only be manufactured by the centreless-grinding process.

Centreless grinding is an alternative manufacturing process that lends itself to the accuracy, production rates and cost control demanded by the high-volume industries. Further, the centreless grinding process is applicable to a wide range of components and a wide range of materials such as ferrous and non-ferrous metals, glass, rigid plastics and wood. Plain grinding and regulating wheels are used for cylindrical components but when using formed grinding and regulating wheels a wide range of components can be made, as shown in Fig. 4.9, where products as diverse as precision-engineering components, wooden draw knobs and glass bottle stoppers can be seen. The wooden articles are ground from the solid in a matter of seconds. The basic principles of the operation of conventional-centreless grinding will now be considered before moving onto the application of CNC control to this process.

Centreless grinding can provide metal removal rates of 0.25 mm per pass when roughing out, yet maintain a dimensional tolerance of 0.005 mm on finishing cuts. The process readily adapts to automatic sizing techniques and also to automatic component feed systems. As the name suggests, the work is not held between centres but is supported on a workrest blade and held up to the face of the grinding wheel by a second abrasive wheel called the control or regulating wheel. This control wheel not only rotates the workpiece but also provides the through-feed in the case of long parallel work. Figure 4.10(a) shows the layout of a typical centreless-grinding machine, and Fig. 4.10(b)

Fig. 4.9 *Examples of centreless form grinding*

shows the relationship between the grinding wheel, the control wheel and the workpiece. Figure 4.10(c) shows the work guide and end stop in plan view. The end stop is not always required. The angularity of the workrest blade is necessary to:

- assist in 'rounding-up' the workpiece and preventing lobing;
- keep the workpiece from being drawn into the grinding wheel and to keep the workpiece in contact with the control wheel;
- provide the control needed to ensure that the workpiece is brought gradually into contact with the grinding wheel when plunge grinding.

The work rest is set so that the axis of the workpiece is slightly above the centre line of the grinding wheel and the control wheel, which helps prevent a form of out-of-roundness called lobing.

4.2.1 *Throughfeed (traverse) grinding*

This technique is used for parallel work of any length such as rods of silver-steel or the rollers of roller bearings. These components are of constant diameter and have no shoulders to prevent them from passing between the grinding wheel and the control wheel. In order to provide axial movement to the workpiece, the control wheel is inclined as shown in Fig. 4.11(a). This inclination produces an axial component velocity and, as is evident from the velocity diagram, Fig. 4.11(b), the greater the inclination the greater will be the feedrate (F).

Fig. 4.10 *External centreless grinding: (a) main features of a typical centreless grinding machine; (b) relationship between work, grinding wheel and control wheel; (c) work guides and end stop*

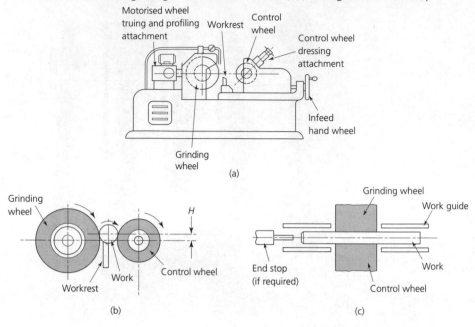

(a)

(b)

(c)

Fig. 4.11 *Control wheel inclination for 'through-feed' grinding: (a) control wheel inclination; (b) velocity triangle*

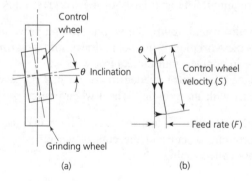

(a)   (b)

4.2.2 *Plunge grinding*

This technique is used for short, shouldered workpieces, multidiameter work, and form-work. It is essential that the length of the component being ground is less than the width of the grinding wheel and the control wheel. Figure 4.12(a) shows some typical workpieces, while Fig. 4.12(b) shows the principle of the technique. An end stop is used to position the work axially and the control wheel is given a slight inclination not exceeding 0.5° to keep the work fed up to the stop.

Fig. 4.12 *Plunge grinding: (a) typical plunge-ground components; (b) set up for plunge grinding*

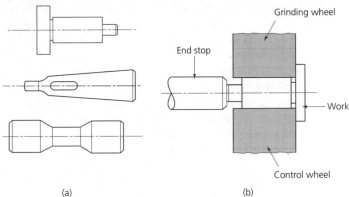

(a)   (b)

After the workpiece has been positioned against the stop, the control wheel is fed forward and advances the rotating workpiece up to the grinding wheel. When the control wheel slide is arrested by a positive stop, the control wheel is allowed to dwell while the grinding wheel 'sparks-out' to leave the workpiece with the correct diameter. The slide and the control wheel are withdrawn and the work is automatically ejected by the end stop. As previously mentioned, the angularity of the workrest blade ensures that the workpiece falls back against the control wheel and prevents it from being drawn into the grinding wheel.

4.2.3 *End-feed grinding*

This is a hybrid technique embodying the principles of both the through feeding and plunge-grinding techniques. It is used for work that is too long for plunge grinding but which cannot be fed through the machine because of a shoulder or other obstruction. Let us now consider the application of CNC control to the centreless-grinding process.

4.3 The development of automated control for the centreless-grinding process

To appreciate the difficulties of applying CNC control to the centreless-grinding process it is necessary first to consider the background to the development of the modern automatic centreless-grinding machine. As described in Section 4.2, centreless grinding has two primary modes of operation, throughfeed and plunge (or infeed) grinding. In throughfeed operation the parts pass through a static grinding gap, advancing as they rotate due to the influence of the inclined regulating-wheel spindle. In plunge-grinding mode the gap is progressively closed upon the component until a fixed position is reached, after which the gap is opened again for the completed part to be removed and the next part to be inserted.

The demands and opportunities for the application of automatic systems are far higher for the plunge operation, because the infeed action of the regulating-wheel slide lends itself to being actuated by mechanical, pneumatic, hydraulic or electrically-driven systems. Unfortunately these relatively expensive systems are largely irrelevant when the machine is used for throughfeed. The throughfeed operation only requires occasional and small incremental movements to maintain the gap, which varies on account of wear occurring predominantly to the grinding wheel; some form of PIC (peripheral integrated controller, also popularly known as a programmable integrated circuit) feed system is therefore all that is required.

Initially, for many reasons, machine tool designers strove to satisfy the requirements of both modes of operation in one machine and the increased cost of accomplishing this was a natural barrier to development of the machine. The increased cost was disproportional to the benefits gained. It must also be remembered that the traditional centreless-grinding machine was a well-tried and tested product with a comparatively long life. Therefore little need was seen for innovation, and the market demand for new and more costly machines flattened dramatically. This brought with it a reluctance to invest in the research and development (R&D) required, and none of the major machine tool manufacturers was interested in introducing new models or concepts. As a result some of the most popular centreless-grinding machines remained in production for as long as 40 years, a factor that on its own suppressed radical development and the introduction of new technologies.

Thus, for centreless grinding, the adoption of full CNC control lagged behind its adoption for other significant machining processes. Early attempts to incorporate electronic control were restricted to logic control systems used to sequence the pneumatically-, hydraulically- and mechanically-actuated mechanisms of the existing generation of machines. As previously stated, in throughfeed grinding there were few worthwhile benefits, but, in plunge grinding there was an increased requirement for automation, particularly with machines that were equipped with pick and place loading and unloading systems. The PLC (programmable logic controller) was increasingly used to replace the former relay logic. This had little impact on the grinding process itself but significantly increased the scope and reliability of specially engineered machines for automatic plunge-grinding operations.

The advent of full CNC control systems awakened an interest in the possibility of automating the centreless-grinding process. However, one crucial obstacle still remained. As with other grinding processes, the stepper motor was found to be difficult to apply reliably to centreless-grinding machines. Two characteristics of the process frustrated the early attempts to use this technology. One was the magnitude of the 'push-off' forces generated by the negative cutting-angle geometry of abrasive grits and the other was the damping qualities required in the slide design to suppress the disturbances created by the millions of tiny 'collisions' between the grit of the abrasive wheel and the workpiece. The stepper motor lacked the power density of a hydraulically-actuated mechanism and this prevented the successful application of a technology that was already revolutionising other machine tool designs and greatly extending their capabilities.

It must be remembered that very considerable forces are involved in plunge grinding. It is not often appreciated that the 'push-off' force has the greater magnitude and is far larger than the shear force required to cut the workpiece material. The grinding wheels that are usually used are 610 mm diameter when new. The width of the face depends upon the length of the job being ground and the increase of closing force with the width of the face of the wheel. For example grinding wheels that are 610 mm diameter with a face width of 250 mm require a power input to their spindle of 40 kW. With the face width increased to 400 mm the power input to the spindle becomes 80 kW. With a maximum width of grinding wheel – currently 1000 mm – the power input to the spindle becomes 200 kW. Also it must be noted that the load on the lead screw feeding the work against the grinding wheel can vary, when plunge grinding, between 22 tonnes when using a 250 mm wide wheel and 130 tonnes when using a 1000 mm wide wheel. With loads of this magnitude it becomes only too clear why the early CNC controlled machines using stepper motors were unsuccessful.

This deficiency was overcome by the adoption of the a.c. servomotor. However, by the time it became available, the market had become understandably sceptical about 'electronic centreless-grinding machines', owing to the lack lustre performance and poor reliability of machines based on earlier pulsed d.c. stepper motor *open-loop* technologies, which lacked feedback.

In the transformation of the traditional hydraulically-controlled machines, the function that benefits most from the potential precision and flexibility of control provided by electronic data processing is the slide axis used for closing the grinding gap. Depending upon the design concept of the machine, this was accomplished by either moving the regulating wheel towards a static grinding-wheel spindle or, in alternative machine arrangements, moving the grinding wheel towards a static regulating-wheel spindle.

As stated previously, the first electronically controlled machines, using stepper motors to actuate the feed mechanism, immediately encountered the limiting factor of insufficient torque to overcome the 'stick–slip' characteristics of slides that had proven damping qualities. Attempts to resolve these problems led different companies in different directions, some sought to reduce stick–slip by using antifriction slideways. This solution proved to be self-defeating because such slideways do not have self-damping qualities. It has since be proved that use of conventional slides actuated by more powerful servomotors is the most satisfactory solution for grinding machines as it damps the vibrations and isolates them from the leadscrews. Other manufacturers sought to increase the mechanical advantage of the stepper motors by the use of gearboxes. The gear ratios available through harmonic gearboxes did not prove to be a satisfactory solution. The adoption of a rotary or linear encoder to provide a *closed-loop system* with feedback, and replacing the stepper motors with more powerful servomotors, enabled the microprocessor system to correct for lost or gained position. This was much more accurate since such a system knew the target position and error. However, it was still difficult to 'jog' to the precise position, even under ideal conditions.

As was stated earlier, centreless grinding results in considerable push-off forces due to the negative cutting-angle geometry of the abrasive grain. Furthermore, the feed force is not constant because the conditions after dressing the wheel quickly alter as the abrasive became dull and less efficient.

An interesting approach to the construction of fully automated centreless-grinding machines at an affordable price has been adopted by *The Centreless Tooling Group*, who 'remanufacture' conventional grinding machines in a significant way. Traditional machines are not only restored to original condition, they are improved and reconfigured to take full advantage of modern CNC control systems. This company has found that the best solution to the application of CNC to centreless grinding lies in combining the self-damping qualities of conventional slideways with powerful a.c. servomotors to over come the stick–slip properties of such slideways. This isolates the vibrations caused by the 'collisions' between the abrasive grit and the workpiece from the leadscrews. The leadscrews themselves are of the recirculating-ball type in the smaller machines and the roller type in the larger machines. The mechanical advantage of the servomotors is further increased by the use of harmonic gear boxes and the traverse screws are located axially by heavy-duty preloaded thrust bearings. The collective process knowledge and applications experience of this company allow its engineers to tailor machines to customers' requirements in order to provide better productivity and improved quality at an economic cost. Inevitably applications arise that demand more than the traditional machines can provide, and this has stimulated research and development into the possibilities of applying CNC in radically new ways to improve the centreless-grinding process.

The technical team engaged in this work was composed of experienced machine tool designers, electronics specialists, programmers, machine tool fitters, run-off and test engineers, service engineers and application specialists. The emerging a.c. servotechnology also offered the opportunity to overcome the shortfalls of the previous stepper motor age, nevertheless it required careful implementation to realise the true benefits.

4.4 The application of CNC to the centreless-grinding process

Computer numerical control can be applied to centreless-grinding machines in many areas, the number of axes used depending upon the application and the demands and expectations of the customer. Fully remanufactured centreless-grinding machines rarely have less than two axis control, one for controlling the grinding gap (the main axis) and the other a rotary axis for turning the regulating (or control) wheel spindle. The rotary axis is relatively inexpensive as a second axis as it immediately avoids the need for gearboxes or other forms of variable-speed drive. The main axis control mechanism comprises a recirculating ball or roller screw and nut, a zero backlash gearbox and an a.c. servomotor.

For less demanding applications 'entry level' machines with just these two axes provide a radical step forward; they are normally supplied with throughfeed and plunge-grinding capability, the mode of operation being selected by a push button. A number of methods are provided for traversing the main slide. The most popular is the use of an electronic hand wheel, which generally has at least two ratios, one for fine movement and a coarse ratio for faster movement that is required when traversing more significant distances for setting or wheel changing. Intermediate ratios can be provided if required, this is simply achieved by programming changes.

Four important functions, actuated by push-buttons, are provided that are used in both modes of operation:

- Fine forward compensation.
- Coarse forward compensation.
- Backward compensation.
- Safety retraction.

On the entry-level machines the least increment of feed is 1 μm, this can be reduced to 0.1 μm on request but at extra cost. Both the forward-acting compensation functions reduce the size of the component while the backward compensation function increases the size. The fine forward functions provide for size control, while the coarse forward function is intended for restoration of the gap after a dressing operation has been completed. The *safety retraction* function provides an emergency retraction by opening the grinding gap. All of these four functions are easily programmed into the controller by keying in the appropriate distances on a menu type screen. The features already described transform the control of a throughfeed operation. Additionally the speed of the control wheel is also variable and can be keyed into the controller. Two speeds are available, one for grinding and one for dressing.

In the plunge-grinding mode, complex infeed cycles can be programmed into the controller. The approach taken for most applications is to provide four phases of easily programmed motion with the addition of a pause at the completion of the stroke that is referred to as 'spark out'. Each phase is programmed using two parameters, distance and speed. The first part of the cycle is the 'rapid approach'; this closes the gap that exists at loading, bringing the component into close proximity with the grinding wheel. In order to give the next phase some relevance it is referred to as 'rough grinding'. This is followed by a 'semi-finishing' phase before the final 'finish-grinding' phase is commenced. These three grinding phases allow some modification of the grinding action to suit the prevailing conditions and provide more efficient cycles than any that would have been available with mechanical or hydraulically-operated infeed motion.

A similar approach is used for other axes; for example two additional axes can be applied relatively inexpensively to the dressing system for the grinding wheel, one for forward incremental movement of the diamond so as to apply a cut, the other to control the speed of traverse across the wheel face. This not only provides a high degree of control and accuracy to the dressing function, it can be used to compose routines far more sophisticated than are possible on traditional systems. By the addition of further hardware the dressing unit can also be prepared so as to impart profiles to the grinding-wheel surface for grinding multidiameter components containing all types of features. A third axis can be used on the dresser to dynamically change the attitude of the diamond as it contours a radius or gradient; with such a system almost any profile can be reliably produced. Similar dressing facilities can be provided for the regulating wheel although it is rare that an application demands the same degree of sophistication as for the grinding wheel.

The Centreless Tooling Group have also applied the control made possible by the a.c. servomotors to other features of the machine, to extend the range of applications and complete features of components that it was not realistically possible to attempt with traditional machines. For example, axial movement of the grinding wheel, the regulating wheel and/or the work support have been used to provide grinding solutions beyond those that were formerly possible.

4.4.1 *Getting slightly smarter*

The features previously described have greatly improved the performance of the machines. However, the application of CNC has exposed possibilities far beyond that of precise control of motion and position. For example, in the plunge-grinding cycle a function referred to as 'diminishing feedrate' has been added to the last phase of the cycle, and this function can be toggled on or off depending on the operator's preference. The diminishing feed facility applies a mathematical algorithm to change the linear speed/ distance profile to a curve similar to simple harmonic motion. The feedrate is constantly reduced as the final position is approached. This allows any pressure that may have accumulated in the earlier part of the cycle to dissipate steadily, 'spark out' is often not required in this mode, resulting in improved finish and repeatability. Certain types of component benefit more from this infeed characteristic than others, for example those with interrupted surfaces (keyways, holes, etc.) or which are out-of-balance.

Another feature closely linked to the diminishing-feed facility is the ability to progressively change the operating speed of the regulating wheel, which is programmed in as a positive or negative integer and changes the speed as a percentage. This allows improved surface textures to be obtained from slightly coarser grinding-wheel compositions that in turn permit more efficient stock removal during the rest of the cycle.

4.4.2 *Getting even smarter*

The use of a.c. servomotors allows more precise positional and speed control, and may be considered as an intelligent 'muscle'. It applies sufficient force to overcome any resistance it encounters in following its instructions. If the force required to achieve its objective is low, the current applied is also low; as greater resistance is encountered more energy (current) is applied and the stiffness of the system can be tuned to complement this function. In principle, a following error is generated by the control, which represents the distance the actual position of the grinding wheel at any instant lags behind the theoretical position. The greater the lag, the greater the current that is applied to keep the following (lagging) error within set parameters. Interrogating these conditions provides a measure of the feed effort being applied. This information can be very useful in monitoring the grinding conditions.

There are two principle forces produced by an abrasive wheel, one relates to the shearing action of the material, the other is the 'push off' force generated by the negative cutting angles. In CNC controlled centreless grinding, the regulating wheel is driven by the servomotor. When a part is not being ground, the energy used is only sufficient to overcome friction and inertia and maintain the programmed speed. When grinding occurs, torque is transferred to the regulating wheel by the component that, in turn, is trying to accelerate due to the shearing forces generated in the material by the grinding wheel. Interrogation of the regulating-wheel servomotor during a plunge-grinding operation would show positive power input trying to rotate the wheel. Thus the power input to the regulating-wheel servo would be reduced as grinding commences, possibly reducing to zero input and even becoming negative (inverted) input, thus acting as a braking action and preventing overspeeding. As the grinding activity reduces towards the end of cycle this power profile will reverse.

This information can optimise the infeed cycle but first let us consider the condition just after dressing. In the early cycles after dressing, the wheel is at its most efficient in removing material. However, as time progresses this efficiency will almost certainly reduce; this changing condition will also be witnessed in the servo drive.

The other force effectively governs the energy requirement of the servomotor driving the infeed axis, and is over and above the force required to move the physical parts of the machine on their slides. This information again provides a useful means of evaluating the cutting characteristics of the grinding wheel. Combining the results from both provides another channel of information.

Power consumption of the grinding wheel can also be monitored, and this provides yet another channel of data that can be used to tune and control the grinding process. The action of a grinding wheel changes as its diameter becomes reduced. Should the grinding-wheel spindle rotate at a constant speed the surface speed of the wheel becomes reduced as the wheel is dressed. This is not particularly significant after one dressing but, over time, a 600 mm diameter wheel may not be discarded until it is 400 mm diameter. This represents a considerable reduction in surface speed. If the grinding conditions are kept the same throughout the life of the wheel, the perform-ance will change appreciably. Modern technology provides a solution in the form of a *constant surface speed* option, the rotational speed of the spindle is automatically calculated by the control unit, and is dependent on the wheel diameter and the surface speed selected by the operator.

Small piezocrystal devices can be attached to, or imbedded in, various locations on the machine, and act like microphones that can 'listen' to the process. Background noise can be electronically removed so that the operator can concentrate on certain frequency ranges, and further tune up, optimise and monitor the grinding process. A piezo-crystal device may be located in the work rest, another in the spindle housing and another in the dressing unit near the diamond. The possibilities for automated fine-tuning are endless.

Vibration can be monitored with an accelerometer, which can be used to control servo-adjusted weights in the wheel mount that balance the wheel, not just initially but constantly throughout its life. Extremely high degrees of balance can be held by this method, leading to better grinding performance and improved component quality in terms of accuracy and surface finish.

Thermocouples in the coolant and machine frame can provide information that can be used to monitor the thermal gradients and remove the effects of thermal expansion on the process. Even the colour and density of the sparks from the wheel can be moni-tored. This is yet another method of monitoring the health of the process.

4.4.3 *Getting really smart*

By gathering information from the machine together with information from the sensors monitoring the machine and the process, the modern centreless-grinding machine will not only produce better quality parts, with higher productivity and efficiency, but it will also convert time lost by the traditional machines into productive grinding effort. Modern CNC machines can become truly *adaptive*, seeking the conditions known to provide the best results, but how is this achieved? A standard CNC device provides a great deal of this information for free, it is the by-product of the control systems used, and other information is relatively inexpensive to acquire, sensors are not expens-ive these days.

However, making good use of the information requires understanding, imagination and intelligent integration of the systems. Most applications are unique and need to be tailor-made to suit individual customer requirements. 'Off the shelf' solutions built into machines on a 'one size suits all' basis will either include a mass of features not required for a given application or, worse, none of the features required, if the bottom line cost dictates the constructional specification. In the case of remanufactured machines, they can be built to reflect the customer's own preferences at minimum cost. A CNC controller reads a sequence of programmed messages (programme blocks) and responds to the commands. For many applications this is more than adequate but for centreless grinding it is too slow.

Let us consider an example of adaptive control. Under a simple CNC control of a plunge-grinding operation, either the control wheel or the grinding wheel will rapid forward to a predetermined point and then slow down to the primary feedrate before executing the operation. The point at which the feedrate changes is at a finite distance from the grinding wheel, for reasons of safety. Thus time is lost when the gap is closing at the primary feedrate before grinding commences. With adaptive control a sensor (or sensors) responds to the closing gap during the rapid forward feed so that, at the moment of contact between the grinding wheel and the workpiece, the CNC program is interrupted and moved to the next block of information concerned with the actual grinding operation. This sensing of the gap closure (gap elimination) and interruption of the CNC program are achieved by external central processor units (CPUs). They receive the messages from the sensors, process the information and interrupt the CNC program in a few nanoseconds. A schematic of this system is shown in Fig. 4.13. In this example, the sensor could be an embedded piezo device that 'listens' to the change in sound as the gap closes and, at a predetermined instant in time, it will send a message to the external CPU that it should interrupt the CNC rapid forward program and change to the primary feed rate.

From time to time in this section the term 'remanufacturing' has been used. This is a technique used by a number of firms manufacturing various types of machines – including *The Centreless Tooling Group* – that starts by taking redundant, conventional machines, virtually at scrap price, thus saving on the cost of expensive patterns and castings. Also, because of the age of the castings they have become dimensionally stable and contain no locked in stresses. The main castings and components are then reclaimed, remachined and reconfigured to accept those control systems and associated sensors that suit the customers' requirements. Furthermore, the original, redundant machine may already be owned by the customer and this reduces the cost still further compared with a new machine built from scratch.

Fig. 4.13 *Adaptive control for a centreless grinding machine*

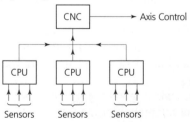

The authors wish to thank Mr Derek Baker, managing director of *The Centreless Tooling Group*, for his assistance in compiling this section of the chapter.

4.5 Turret punching

Initially, CNC control was developed for, and applied to, the engineering processes of turning, milling and grinding. However, the benefits of CNC control proved to be so great that CNC control was soon adopted across a much wider range of processes and industries. This is why the principles of CNC programming and the more important options available to the programmer have been dealt with in some detail in the previous sections of this chapter. Let us now consider the application of CNC control to manufacturing processes other than turning, milling and grinding.

4.5.1 *Sheet metal stamping (principles)*

The blanking, piercing and forming of sheet metal by pressing (stamping) is considered in some detail in *Manufacturing Technology*, Volumes 1 and 2. Figure 4.14(a) shows the basic principle of stamping a hole (piercing) in sheet metal, while Fig. 4.14(b) shows a simple piercing tool for stamping holes in sheet metal components. The downward force on the punch shears the stock metal leaving a waste slug that drops through the die. To avoid undue wear on the tools and to ensure a clean cut, there must be a suitable clearance between the punch and the die as stated in Table 4.1. Note that the punch is made the size of the desired hole and the die is made slightly larger.

One of the problems of conventional pressing is the high cost of the tools required and the extended lead-time in their manufacture. Where large-volume production, such as body panels for the automotive industry, is concerned this is not a great problem since the tooling costs can be recovered. However, when relatively short-run production is required together with frequent changes in the product being made, it is difficult to recover the cost of dedicated tooling. One solution is the CNC controlled Turret Press as shown in Fig. 4.15.

The tooling is located in indexable turrets. The upper turret carries the punches and the lower turret carries the dies. These are selected automatically under CNC control to suit the job in hand. The X- and Y-axis movement of the sheet metal blank between the

Table 4.1 *Die clearances*	
Material	Die clearance per side (double value given for diameters)
Aluminium	$1/60$ material thickness
Brass	$1/40$ material thickness
Copper	$1/50$ material thickness
Steel	$1/20$ material thickness

Fig. 4.14 *Hole punching in sheet metal: (a) piercing action; (b) piercing tool*

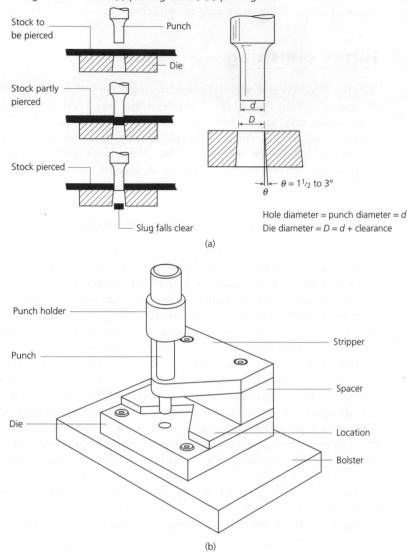

Stock to be pierced
Punch
Die
Stock partly pierced
Stock pierced
Slug falls clear

d
D
$\theta = 1^{1}/_{2}$ to $3°$
θ

Hole diameter = punch diameter = d
Die diameter = $D = d$ + clearance

(a)

Punch holder
Stripper
Punch
Spacer
Die
Location
Bolster

(b)

tool turrets is also under CNC control. The machine movements are actuated hydrauli-cally under CNC control. Hydraulic clamping is also used and the raw material (sheet steel) is gripped 5 mm from one edge to minimise waste. Let us now consider the fea-tures of a typical machine such as the *Amada Vipros King* as shown in Fig. 4.15.

4.5.2 *The Amada Vipros turret punching machine*

In this machine there is an 'intelligent' hydraulic ram system that ensures that the optimum ram cycle is used for forming, nibbling (the production of an enlarged hole or

Fig. 4.15 *CNC controlled turret punching press*

a slot by piercing a sequence of overlapping holes), slitting and marking. The machine will also teach itself the optimum ram movement that will optimise productivity and minimise noise. The machine also has a 'brush table' design that not only further reduces noise but also ensures that the surface of the sheet metal is not scratched, thus eliminating secondary finishing. The triple track turret allows for quick track to track tool changing and has 58 stations, of which 4 are auto-indexing. The two $4^1/_2$ inches auto-index stations allow a range of special shapes to be rotated in 0.01° increments. The multitool adaptor using standard tooling minimises set-up time.

Punching is carried out at high speed, achieving a maximum of 1200 hits per minute and maintaining 460 hits per minute on a 25 mm pitch. This is achieved through the use of an intelligent servocontrolled hydraulic ram. At 113 m min^{-1} the machine table and, therefore, the workpiece are positioned at high speed under the turret. Obviously considerable heat is generated when operating at such high speeds and the tooling is cooled by an air-blow/oil-mist system to ensure long tool-life. The air-blow system accurately injects a combination lubricating/cooling oil and compressed air (oil mist) into the air-blow tool. The oil mist is distributed through the punch body into the guide, out to the turret bore and down onto the workpiece just prior to punching.

This effective combination ensures that the tool is lubricated in all essential and critical areas. In addition, heat generated during punching is dissipated away through the oil mist. This solves many problems familiar to conventional tooling resulting in up to a three-fold increase in tool life. This system is particularly effective in overcoming the phenomenon known as *microfusing*. This is the cold-welding of shards of the component material to the punch and die cutting edges, which causes rapid wear.

As the oil mist is pressurised, powdery substances created during punching, along with fine 'needles' and any loose slugs, are ejected harmlessly through the die ensuring that both the tooling and the workpiece are kept clean as shown in Fig. 4.16(b). This results in improved component quality and dramatically reduces tooling maintenance. Because the air-blow system is operated by the program it can be directed to the most critical areas, where it will have the maximum effect.

Fig. 4.16 *Air-blow tooling: (a) air-blow mist lubrication – lubricant supply;*

(a)

A typical tool is shown in Fig. 4.17. The slugs from the punching operations drop through the hollow dies into a hopper for disposal. The machine ram (striker) strikes the head of the tool, pressing the punch through the metal so that the unwanted slug falls through the die aperture. At the same time the guides retract against a powerful spring. As the machine ram rises, ready for the next blow, the spring retracts the punch and strips it from the sheet metal, thus the guide acts as a support to the punch and as a stripper. Typical standard punch and die profiles are shown in Fig. 4.18.

Variable dwell time and position at the bottom of the stroke provides high quality forming. Progressive forms, flanges and embossing eliminate the need for secondary processing. For instance, louvres can be raised, parts can be progressively numbered with the number increasing incrementally for each component automatically and burred holes can be punched to provide extra thickness for tapping screw threads. Punching and tapping can be combined under CNC control. The high-speed tapping attachment can be inserted in the turret where it converts the linear stroke of the striker into rotary motion for tap rotation. It is fully compatible with punch press performance in terms of speed, accuracy and repeatability.

After its profile has been 'nibbled around', the component is left attached to the skeleton of the sheet by microwebs at its corners. These microwebs are only 0.1 mm × 0.1 mm but enable the skeleton of the sheet and the required component to be easily removed from the machine onto the *shake-out* table by the operator. While the next component is being made, the operator literally shakes the required component out of the surrounding skeleton. The scrap skeleton is folded up and discarded. Figure 4.19(a) shows the layout for a typical 'standalone' machine and Fig. 4.19(b) shows a typical component produced on a turret punching machine.

Fig. 4.16 *(b) oil mist passage through the punch*

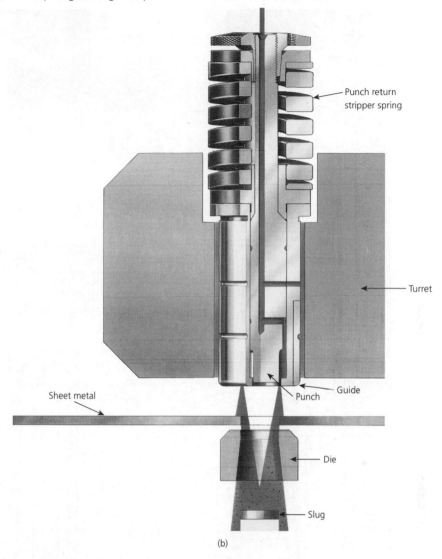

Punch return stripper spring

Turret

Sheet metal

Guide

Punch

Die

Slug

(b)

4.5.3 *Automated turret punching*

Modular automation is also possible in order to meet the requirements for higher output, reduced costs and unmanned operation. Sheet load/unload systems are available in four sizes up to 3000 mm × 1500 mm and thicknesses from 0.5 mm up to 6.4 mm. With 6 tonnes raw material capacity the automatic pallet change can sustain long, uninterrupted production runs. To complement the sheet load/unload module, the part remover is capable of picking up single parts from 100 mm × 150 mm up to 1500 mm long or 1000 mm wide. The automation systems include an integral guarding system.

Fig. 4.17 *Typical turret punch press tools*

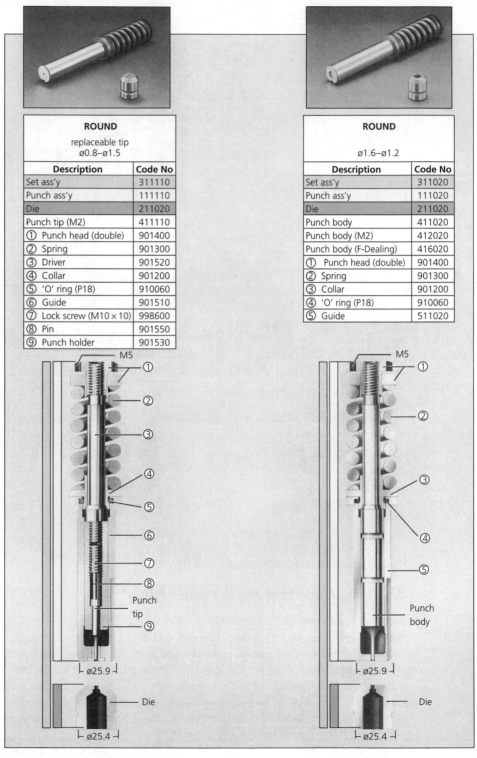

ROUND replaceable tip ø0.8–ø1.5	
Description	**Code No**
Set ass'y	311110
Punch ass'y	111110
Die	211020
Punch tip (M2)	411110
① Punch head (double)	901400
② Spring	901300
③ Driver	901520
④ Collar	901200
⑤ 'O' ring (P18)	910060
⑥ Guide	901510
⑦ Lock screw (M10 × 10)	998600
⑧ Pin	901550
⑨ Punch holder	901530

ROUND ø1.6–ø1.2	
Description	**Code No**
Set ass'y	311020
Punch ass'y	111020
Die	211020
Punch body	411020
Punch body (M2)	412020
Punch body (F-Dealing)	416020
① Punch head (double)	901400
② Spring	901300
③ Collar	901200
④ 'O' ring (P18)	910060
⑤ Guide	511020

Fig. 4.18 *Standard turret punch tool shapes with key angle and clamp positions*

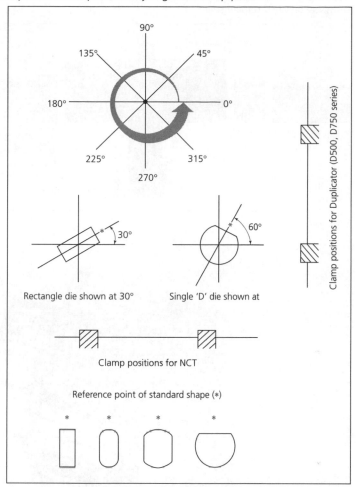

The automation modules can easily be retrofitted to the basic machine and range from simple load/unload units to full flexible manufacturing systems (FMSs).

4.5.4 *Load/unload cell*

The layout of a load/unload cell is shown in Fig. 4.20. The cell is under the control of the CNC unit fitted to the turret punching press. This sends a signal to the loader when loading is required and an 'unload' signal to the unloader when the operation is complete. Once the 'load' signal has been sent to the loader, its own processor takes over and sequences the loading operation. Similarly, when the punching machine has finished producing the component, the CNC controller sends a signal to the unloader and the unloader's own processor takes over and sequences the unloading operation. The component is held in the skeleton by *microwebs* and is removed by hand on the 'shake-out' table while the next component is being made.

Fig. 4.19 *Typical stand-alone layout and punched component: (a) stand-alone machine; (b) typical punched component*

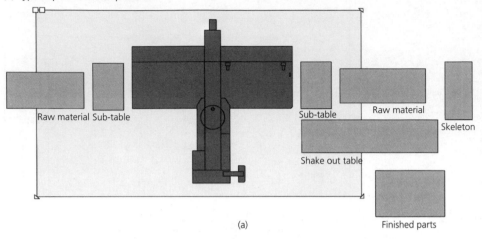

(a)

(b)

Fig. 4.20 *Load/unload cell*

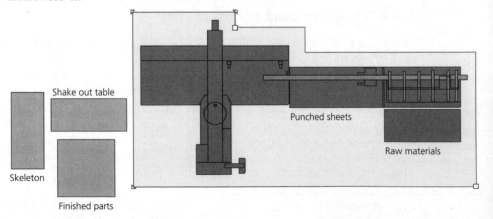

Fig. 4.21 *Load/unload cell with automatic part removal*

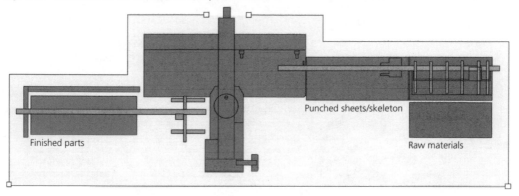

4.5.5 *Load/unload cell with automatic part removal*

The layout for a load/unload cell with automatic part removal is shown in Fig. 4.21. This is used where a number of components (parts) are being manufactured from one sheet. The sheet loading sequence is the same as described previously. However, this time, the CNC control unit stops the process immediately before the last cut that would normally free the first component from the skeleton is made. This ensures the component is retained in position. The controller then sends the unload signal to the *part removal unit*, which then grips the part ready for removal. The controller then instructs the turret punching machine to make the last cut and as soon as the last cut is made the part removal unit's own processor takes over and removes and stacks the component. The CNC control unit then instructs the punching machine to make the next part and the sequence for unloading is again carried out by the part removal unit. This is repeated until the last part is removed from the sheet. The CNC control unit then instructs the loader/unloader to remove the scrap skeleton and load the next sheet. The part removal unit stacks the parts alternately at right angles to each other for ease of subsequent handling.

Thus by saving tooling costs by the use of standardised tooling, and giving high rates of production, flexibility of manufacture, automation and a high quality product, the turret punching process meets most of the parameters set out in Section 1.7.

4.6 The hydroforming process

Another interesting development in metal forming has been the substitution of many conventionally pressed (stamped) components for the automotive industry by *tube hydroforming*. Again this is a process made possible by the advent of electronic data processing. The following information on the *VARI-FORM advanced hydroforming* process is based on data provided by Mr Derek Payne of VARI-FORM Inc., Canada. The design of the product and tools by CAD and the preforming of the tube in CNC controlled bending machines are very important parts of the process. Also if the *VARI-FORM advanced hydroforming* process is used, the tube fluid pressure must be carefully controlled and timed to occur at the correct stages of die closure. This is best

Fig. 4.22 *Hydroforming applications*

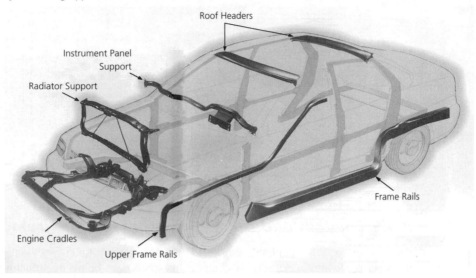

achieved by the use of programmable logic controllers linking the tube pressure and the press cycle. The selection and pre-forming of the tube is often the limiting factor in the process and will be considered in Section 4.6.3.

Tube hydroforming is a pressurised hydraulic forming process in which preformed metal tubes, circular or elliptical in cross-section, are inflated under very high fluid pressure (as much as 40 000 lbft in^{-1} (275.8 MPa)) so that they take the form of the die in which they are constrained. Components produced by tube hydroforming are lighter, stronger and require fewer pieces than traditional components fabricated from welded steel pressings (stampings). Hydroformed components not only retain their structural integrity, but significant cost savings are achieved in lower material useage, lower tooling costs and lower labour costs. The process is applicable to any industry where complex shapes must be formed with a high degree of precision. For example, automotive applications include radiator enclosures, space frame, frame rails, engine cradles and other subassemblies, as shown in Fig. 4.22. It should be noted that in the case of the radiator support shown, a 10-piece hydroformed assembly replaces a 17-piece conventionally stamped assembly, resulting in a 5 kg saving in weight. The hydroformed assembly is prefabricated from galvanised and annealed steel tube, with over 80 punched access and mounting holes.

4.6.1 *Advanced hydroforming technology for complex components*

The VARI-FORM Company has enhanced the capabilities of traditional hydroforming. Their patented *sequential pressurisation* process facilitates the manufacture of highly complex components featuring inside corner radii as low as twice the metal thickness, and intricate bends as well as expanded tube ends.

Fig. 4.23 *High pressure hydroforming – corner forming mechanism*

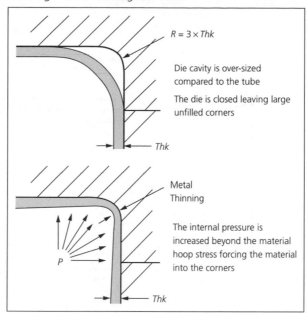

In traditional high-pressure hydroforming, the internal tube pressure is increased until the hoop stress at the corner radius of the cross-section is greater than the yield strength of the material. This results in metal thinning and loss of structural integrity as the tube fills the corner radius of the die by local stretching of the tube wall as shown in Fig. 4.23.

Pressure sequence hydroforming, however, overcomes this disadvantage by forcing the tubular blank to flow into the corner areas of the die without stretching or expanding the tube to fill the die cavity. This is achieved by using a low pre-pressure hydraulic fluid, as shown in Fig. 4.24, so as to exceed the yield limit of the tube material in a bending mode and form the corner radius area while the dies are closing. The pressure is then increased to set the accuracy of the final form and to provide back-up for punching. This is contrary to traditional high-pressure hydroforming systems that fill the corners in tensile mode. Since the tube forming is controlled as the die is closing, the tendency for the tube to 'pinch' at the die split line, as is common with traditional high-pressure hydroforming, is eliminated.

VARI-FORM hole-piercing technology allows components to be designed with any number or shape of holes that can be produced 'in-the-die'. In conventional stamping processes a punch and a die are required to pierce a hole. In the hydroforming process, only a punch is required since the high-pressure water provides the back-up support normally provided by a die. Manufacturing (tooling) holes, clearance holes, and self-tapping or fastener holes can all be accommodated. Such holes can be clean pierced or extruded for a longer support collar or threaded land. Holes can be round, hexagonal, square, rectangular, oval, or irregularly shaped. Unwanted slugs can be removed through the larger holes. Also, VARI-FORM's technology ensures that deformation around the holes is minimised.

Fig. 4.24 *Pressure sequence – corner forming mechanism*

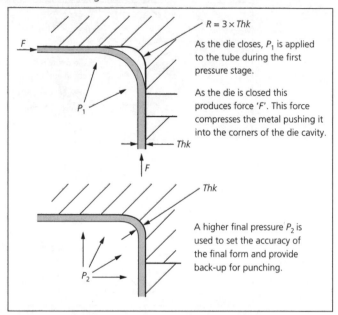

$R = 3 \times Thk$

As the die closes, P_1 is applied to the tube during the first pressure stage.

As the die is closed this produces force 'F'. This force compresses the metal pushing it into the corners of the die cavity.

Thk

Thk

A higher final pressure P_2 is used to set the accuracy of the final form and provide back-up for punching.

4.6.2 *Manufacture by the advanced hydroforming process*

The VARI-FORM process is dependent upon e-manufacture for CNC tube preforming, the CAD design of the product and the forming dies using CATIA software for its success and PLC control of the forming cycle. The tooling required for this process can be divided into three areas:

- pre-operation processes such as tube bending and mechanical pre-forming;
- hydroforming dies;
- post-operation processes such as shearing, coining and welding.

Post-hydroforming and inspection processes are also dependent on computerised manufacturing techniques. The stages in a typical hydroforming process are as follows:

- *Specify tube blank size and composition.* Tube blanks typically range from 1 metre to 3 metres in length and 25 mm to 150 mm in diameter. However, unlike conventional hydroforming, the VARI-FORM process allows tube blanks to be manufactured from a variety of less expensive tube materials including low-carbon hot-rolled steel, cold-rolled steel, high-strength low-alloy steels, pre-coated steels such as galvaneal and even aluminium. Furthermore, changing the material gauge does not require costly tooling revisions as it does in conventional stamped assemblies.
- *Pre-bend blank to approximate configuration.* To prepare the tubular blank for the hydroforming process, the tube blank is bent to approximately the final shape on a CNC tube bending machine as discussed in Section 4.6. Unlike conventional hydroforming,

Fig. 4.25 *Pre-formed tube*

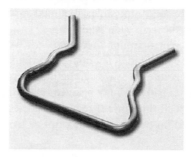

Fig. 4.26 *Sequential pressure forming: (a) first stage – low pressure; (b) second stage – high pressure*

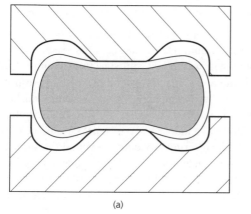 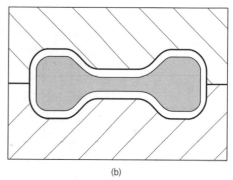

(a) (b)

VARI-FORM blanks require no expensive annealing or lubrication prior to hydro-forming. An example of a pre-formed tubular blank is shown in Fig. 4.25.

- *Place the prepared blank in the forming die.* The pre-formed tubular blank is placed in a die built to the dimensions of the final component.
- *Sequential pressurisation.* This process controls and varies the water pressure and die forces under PLC control. As the dies are closing, the blank is pressurised to assist tube forming. Low pressure at this stage allows the metal to slide along the die surfaces and into the corners, as shown in Fig. 4.26(a). This ultimately ensures uniform wall thickness and accurate calibration. After the dies are closed, the fluid pressure is increased completely forming the sides of the tube as shown in Fig. 4.26(b). Overall, the sequential pressurisation process significantly lowers the forming pressures compared with conventional processes that rely solely on a single-stage high-pressure regime. This results in shorter forming cycle times and requires a smaller and less expensive press. Furthermore, the need for additional steps such as annealing, lubrication and washing is eliminated.
- *Punch access and mounting holes.* During the final pressurisation, punches mounted in the forming die are used to pierce holes in the tube being formed. This 'in-the-die' process minimises stretching and other deformation mechanisms that can adversely affect the final quality. The VARI-FORM process is capable of piercing

a large quantity of holes in a single component in such a wide variety of shapes and sizes. In addition, slides mounted in the die can be used to provide indented surfaces on the tube side walls.

- *Third stage pressurisation to expand ends*. When a component needs peripheral expansion to increase load/deflection performance, or to reduce resonant frequencies, a third intermediate pressure can be incorporated into the hydroforming process. This 'in-the-die' expansion may be accompanied by an end push on the tube resulting in minimised wall thinning. If reduced end sections are required, a mechanical process is performed on the round tube prior to hydroforming.

4.6.3 *Performance*

Let us now consider some of the main performance goals when comparing hydroforming with conventional pressing and how they may influence product design.

- *Deflection under load*. To minimise the deflection measured on a hydroformed tube placed under load, the designer should use the largest cross-sectional area possible. Large thin-wall cross-sections will weigh less and structurally outperform a smaller heavier walled section. Unfortunately, specifying the use of large thin-wall sections can also affect other factors such as fastener retention, welding and tube cost.

- *Crash absorption*. In order to meet all of the criteria of crash absorption in the case of components used in the automotive industry, a part must be capable of withstanding a certain amount of load without suffering any damage yet, at the same time, it must be able to yield and absorb significant amounts of energy when a large load is applied.

- *Torsional rigidity*. Round tube is the structural shape most efficient in withstanding a torsional load. Hydroformed tube vehicle structures have proven themselves, on a weight for weight basis, to be far superior to conventional structures in resisting torsional bending.

- *Material stress level*. Similarly to deflection under load, the material stress level will be minimised most efficiently through the use of large tube cross-sections. Here the use of high strength materials will maximise the load carrying capability of the component while reducing cost and weight. All materials that are strained tend to become stronger. This phenomenon is often referred to as strain-hardening or work-hardening. Caution must be exercised when using this increased yield strength as a basis for stress analysis because the effect will be lost in the immediate vicinity of any welding, owing to local annealing. Wall thickness is also sacrificed when the material is intentionally work-hardened through expansion as is done in traditional hydroforming. Expansion of the periphery has also been shown to cause necking failures in the material of the tube wall that cannot be detected by the hydroforming process. This is largely avoided when the VARI-FORM pressure sequence hydroforming process is used.

- *Weight*. Hydroforming can offer significant weight advantages compared to existing stamped (pressed) and welded assemblies. The elimination of spot welding flanges can result in a 20 per cent reduction in the weight of welded assemblies. Pressure sequence hydroforming has proven itself to be compatible with many alternative materials that can offer even more significant weight savings, though at a cost penalty compared to steel.

- *Noise, vibration and harshness* (NVH). These characteristics are a critical measure of the success of a component. The rigid closed cross-section and continuous weld along the length of a hydroformed tube provide it with natural NVH characteristics far superior to those traditionally found with stamped and welded assemblies.
- *Dimensional tolerances.* All processes will create parts that will vary within a natural range. If the design of a product is such that the natural variation of the product does not negatively influence the quality of the final part then specifying a tighter tolerance will not result in a better part. Producing the component to tighter tolerances will only add cost without providing any added benefit. For this reason tolerances should only be specified where needed and then only as tight as is required to produce a functional part.

Let us now consider some of the design features that could affect cost or affect performance, together with their associated processes.

- *Bent tube.* The tube must be cut to length and bent to the approximate centreline of the finished part to enable the tube to be placed in the die cavity prior to hydroforming. During the bending process the tube undergoes significant coldworking. Often it is the bending process rather than the hydroforming operation that establishes the minimum allowable material-formability properties.
- *Tube cross-section.* Hydroforming can create almost any cross-section shape desired. In cases where the required final shape is significantly narrower in plan view than the diameter of the start tube, a preform operation may be required prior to hydroforming. This will allow the starting tube to fit into the die cavity.
- *Holes.* Piercing while the component is in the die always offers the most economical and dimensionally stable method of providing holes or openings in a hydroformed component. It is estimated that in 2002 over 100 million holes per year were formed using sequential pressure hydroforming. Adding piercing to the forming operation can add some tool costs to the hydroforming die but will not significantly affect capital, labour or operating costs as these remain relatively unchanged with or without piercing. Combining forming and piercing into a single operation ensures that the holes will always be punched in the correct orientation to the part surface and to one another since no secondary operation is required.

4.7 Tube bending under CNC control

Tube bending under CNC control has applications in a wide range of industries that undertake the volume production of preformed pipe. For example oil pipes for the lubrication systems of machine tools, automobile exhaust systems and the preforms used in the hydroforming process. Many tube types can be used when hydroforming but the type of tube most widely used in the Vari-Form hydroforming process is low-carbon, electric resistance welded steel tube (ERW).

As described in Section 4.6.2, when producing hydroformed components for automotive and other applications, the tube that will become the finished part usually requires a bending operation to form it to the approximate shape of the die recess prior to hydroforming. While significant consideration is often given to the capabilities of the hydroforming process and cost of the related dies when designing the part shape, the limitations and costs associated with bending technology are often ignored. This can lead to part designs that require unnecessarily complicated and, thus, expensive bending operations prior to hydroforming. By giving proper consideration to the problems likely to be met during bending at an early stage in the product design process, significant savings can be realised in both up-front capital costs and ongoing process costs.

4.7.1 *Bent tube geometry definitions*

The two common formats used for defining the geometry of bent tubes are XYZ (Cartesian coordinates), and LRA (length, rotation and angle). The latter is also known as YBC, which denotes the axis designations for the corresponding motions of the bending machine.

XYZ format

The XYZ coordinate system is defined by three mutually perpendicular axes that correspond to three dimensions in space. The tube geometry is defined in this system by the location of the intersection points of lines along the tube centreline, as shown in Fig. 4.27. The series of vectors joining these points, along with two additional points at the ends of the tube, define the path the part takes in space. The centreline bend radius at each intersection point completes the part definition.

When using this format, care must be taken when defining a part with a 180° bend because the two vectors surrounding the bend are parallel and therefore the intersection point becomes undefined. Common practice is to break any bend over 175° into two smaller angles with an additional point placed in the centre of the bend to avoid this problem.

Fig. 4.27 *Tube shape in XYZ coordinate system*

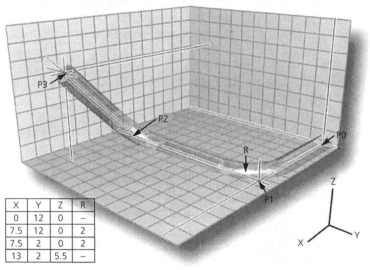

X	Y	Z	R
0	12	0	–
7.5	12	0	2
7.5	2	0	2
13	2	5.5	–

Fig. 4.28 *Tube shape with 'normalised' vector set*

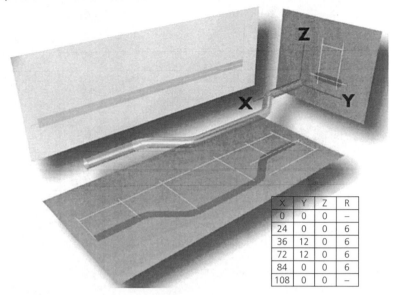

X	Y	Z	R
0	0	0	–
24	0	0	6
36	12	0	6
72	12	0	6
84	0	0	6
108	0	0	–

When using the XYZ geometry definition, the coordinate axes reference position can be chosen in several ways. The so-called 'normalised' coordinates have the first end point of the tube at the origin point, the first straight along the positive X-direction and the first bend in the positive X–Y plane as shown in Fig. 4.28.

Alternatively, for automotive parts, the component is often defined in terms of the 'car coordinates' where the axes are located and oriented at a fixed position on the car. In this case the component is defined by the location it will actually occupy when installed in the car as shown in Fig. 4.29.

When hydroforming, a third useful coordinate system can be defined using the press or die as the reference. In this case, the part is defined relative to a fixed point on the forming die as shown in Fig. 4.30. This system may be the same as the car coordinates depending upon the design of the die but often this is not so. Note that the arbitrary selection of the reference point for this type of geometry definition means that many apparently different sets of points can be used to define the exact shape of the part.

Length, rotation, angle format

A second format often used to define the shape of a bent tube is by the lengths, rotations and angles that actually make up the part. In the industry this is often abbreviated to LRA.

The *length* is defined as the straight distance from the end of one bend to the beginning of the next bend, as shown in Fig. 4.31(a). This is also referred to as the 'distance between bends' (DBB).

The *rotation* is the angular rotation required between the plane of one bend and the plane of the next bend, as shown in Figure 4.31(b). This is also referred to as the 'plane of bend' (POB).

Finally, the *angle* is the angle of the actual bend, as shown in Fig. 4.31(c). This is also referred to as the 'degree of bend' (DOB).

Fig. 4.29 *Tube shape in car coordinate system*

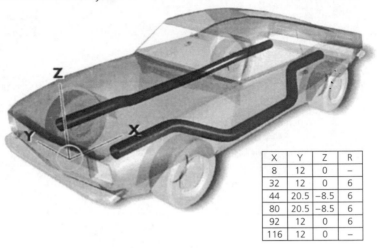

X	Y	Z	R
8	12	0	–
32	12	0	6
44	20.5	–8.5	6
80	20.5	–8.5	6
92	12	0	6
116	12	0	–

Fig. 4.30 *Tube shape in 'die-coordinate' system*

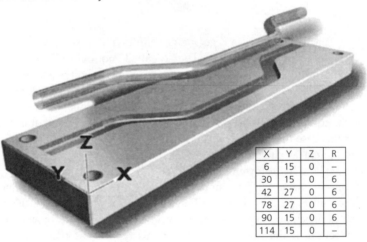

X	Y	Z	R
6	15	0	–
30	15	0	6
42	27	0	6
78	27	0	6
90	15	0	6
114	15	0	–

4.7.2 *Data format conversion*

The XYZ format for the geometry definition has the advantage of being easily extracted from CAD drawings. Thus, this is the system typically used by engineers when designing parts and the necessary data is often located on the part detail drawings. By contrast, the LRA format is not usually found on the part detail drawings because it is defined around the manufacturing process only and does not relate to how the part was designed. The LRA data must therefore be calculated based on the XYZ print data. While this is a relatively simple mathematical conversion, care must be taken when converting between XYZ and LRA data sets to observe the projection symbol on the part drawing. To confuse first- and third-angle projection would result in a mirror image part.

Fig. 4.31 *(a) Length, (b) rotation and (c) angle*

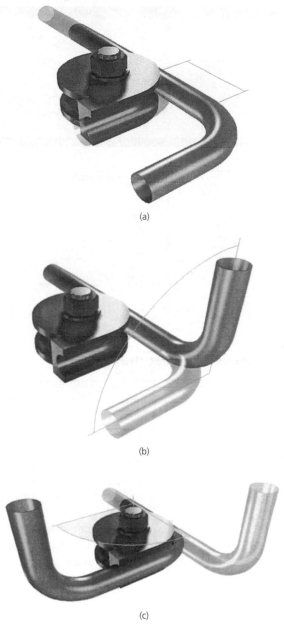

(a)

(b)

(c)

One of the potential difficulties of using the XYZ data format is that the data definition is completely independent from the process used to form the part. In addition, most people find it very difficult (if not impossible) to visualise the bending motions required to produce a given part shape based on the XYZ data. Consequently, when parts are designed working only with XYZ data format, the resulting designs are often unnecessarily difficult and, therefore, costly to manufacture.

Fig. 4.32 *Geometry correction example – point 3 is off by 1.0" in Y direction*

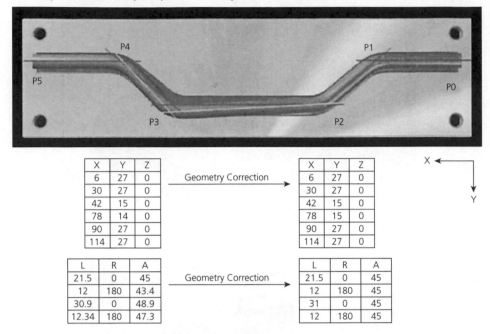

X	Y	Z
6	27	0
30	27	0
42	15	0
78	14	0
90	27	0
114	27	0

Geometry Correction →

X	Y	Z
6	27	0
30	27	0
42	15	0
78	15	0
90	27	0
114	27	0

L	R	A
21.5	0	45
12	180	43.4
30.9	0	48.9
12.34	180	47.3

Geometry Correction →

L	R	A
21.5	0	45
12	180	45
31	0	45
12	180	45

4.7.3 *Correcting geometry*

The XYZ geometry is generally the most useful format to use when attempting to fine tune a bent part shape to fit into the die. Using the XYZ data to correct the part shape relative to the forming die is straightforward because any point on the tube can be moved by a known amount without affecting the rest of the geometry, as shown in Fig. 4.32(a). If the same correction is made using the LRA data, the changes required are much less intuitive. This is illustrated by the updated LRA data in Fig. 4.32(b). To accomplish the same simple geometry correction requires modification to three bends and two straight lengths. This becomes significantly more complex when the part does not lie in a single plane.

In addition, if an early change is made to a bend in the part in an attempt to correct the geometry, all the subsequent bends end up in new locations relative to the die. That is, other changes are also made to compensate for the first change, as shown in Fig. 4.33. This makes it very difficult to isolate changes in one area of the tube. By using the data format conversion included in modern CNC benders, the complications of correcting geometry using the LRA format can be avoided.

4.7.4 *Principles of rotary-draw bending*

The most common type of CNC controlled bending process is *rotary-draw bending*. This process is basically a metal drawing operation where the tube is wrapped around a radius block to form the required bend(s). The principles of the rotary-draw bending process and the tooling required are as follows.

Fig. 4.33 *Geometry correction using angles. Correction to angle at P2 moves remainder of tube out of the die unless additional corrections are made*

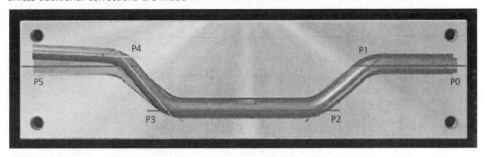

4.7.5 *Basic tooling components*

There are five basic tooling components used in rotary-draw bending operations. The typical tooling set-up is shown in Fig. 4.34.

The bend die

The *bend die* (radius block) provides the radius about which the tube is formed. The outside radius of the die has the profile of half the tube cross-section, to contain the tube during bending. The radius block also includes a straight length, tangential to the radius, to provide an area in which the tube can be gripped for bending.

The clamp block

The *clamp block* mates with the straight clamping area on the bend die and holds the tube in position during bending. Both the bend die and the arm of the bender are mounted

Fig. 4.34 *Typical bender tooling set*

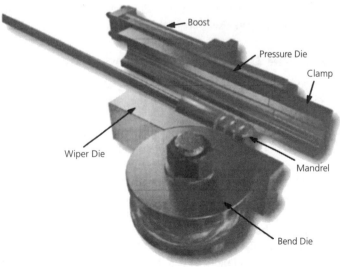

on another arm that rotates about the centre of the bend die. It is this rotary action that pulls the tube to form the bend around the bend die.

The pressure die

The *pressure die* (also known as the reaction block and/or follower die) is located so as to provide the reaction force to the bending torque. When the bend die and clamp rotate with the tube clamped in place, the natural tendency of the tube is to remain straight and simply rotate with the die. The reaction block contains the trailing end of the tube and keeps it aligned with the bed of the machine thus forcing the tube to bend around the radius block. A second function of the pressure die is to provide forward boost to the tube during bending. This is typically accomplished using a hydraulic cylinder that pushes the die forward during bending thus helping to move material along the outer wall of the tube into the bend area.

The bend die, clamp and pressure die, as have been described, are required in all rotary draw bending applications. Under certain conditions, depending on the tube diameter, wall thickness and bend radius, two additional tooling pieces may be required to provide additional support for the material during the bend.

The mandrel

During the bend, the material along the outer wall is under a considerable tensile load that can lead to excessive flattening of the outer wall of the tube unless the material is supported from the inside. This internal support is provided by a *mandrel* that consists of a straight shank portion and one or more balls connected by flexible links or ball joints. This mandrel is slightly smaller than the inside diameter of the tube and prevents the tube from collapsing during bending.

The wiper die

While the outer wall is under tensile load, the material along the inside of the bend is subjected to high compressive forces. If the compressive force exceeds the column strength of the material, the inside wall tends to buckle leading to wrinkles along the inside of the bend. To prevent this, a *wiper die* is added that, in conjunction with the mandrel inside the tube, gives support to the inner wall.

4.7.6 *The basic tube-forming process*

The tube-bending sequence begins with the tube being advanced along its axis and rotated, to position it where the bend needs to be formed. Once in place, the clamp closes against the corresponding clamping area on the bend die so as to grasp the tube and the pressure die closes on the tube to keep the trailing end of the tube aligned with the machine. If a mandrel is being used, the mandrel also advances to move the flexible ball section into the area of the tube being formed. With all of the tooling in place, the bend die and clamp are rotated by the desired amount pulling the tube along with them. At the same time, the pressure die is boosted forward to help move material into the bend to reduce wall thinning. The sequence of operations is shown in Fig. 4.35.

When the bend is complete, the mandrel is retracted back out of the bend area. This motion reforms the material to some degree where the outer wall has collapsed slightly between the mandrel balls. When the mandrel is clear, the clamp and pressure die release

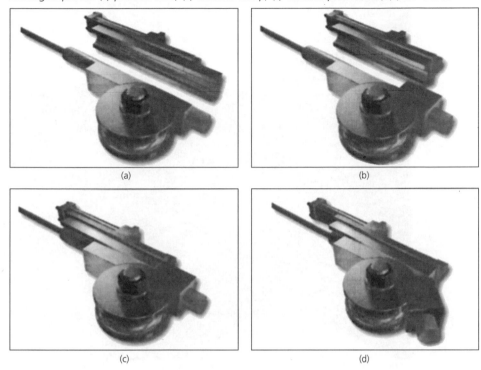

(a) (b)

(c) (d)

the tube and it can be indexed into the position for the next bend. The bending process reduces the cross-sectional area of the tube in the bend area. This effect becomes more pronounced as the bend radius becomes tighter. This factor is also affected by the size of the mandrel relative to the inside diameter of the tube at the start, since the tube will collapse down to the size of the mandrel during bending. Finally, allowance must be made for *spring back*. This is usually greater for large bend radii and high-strength materials and must be compensated for by *over-bending*.

4.7.7 *Basic machine considerations*

A basic rotary-draw bending machine consists of a transport carriage, a bend arm and a reaction arm all mounted onto a fixed machine base, as shown in Fig. 4.36. The transport carriage holds one end of the tube in a collet chuck and provides the linear and rotational transport motions between bends to position the tube for each bend. The bend head is located at the front of the machine and contains the mounting areas and actuators for the bending tools.

Most benders used in high-volume production are CNC controlled machines. The three controlled axes are the length, rotation and bend angle; the LRA data that we discussed earlier under basic tube geometry definitions. Two additional axes can be added to the basic machine that allow for shifting the tube between multiple bend dies on the same machine. This is required for some parts for which the geometry requires multiple

Fig. 4.36 *Typical three-axis bending machine*

bend radii and/or contoured clamp sections. Contoured tooling is shown in Fig. 4.37. The tube is bent round the lower die, retracted, and then the next bend is taken around the upper die, which has a different radius and contour. These operations are all under CNC control and these multiple dies are referred to as stacked tooling.

One final option that is popular in high volume applications is an automatic loading and unloading system. Loading systems can be configured to take straight tubes directly from a hopper, feed them into a weld-seam orientation station and then load the tubes directly into the bender, as shown in Fig. 4.38. The bent tubes can be unloaded with a simple pick and place mechanism controlled by a PLC acting as a slave to the machine CNC unit, as described in Chapter 5. Alternatively a programmable robot can be used, also acting as a slave to the machine CNC unit, depending upon the requirements of the subsequent operations. When designing the loading and unloading systems, the envelope required by the tube as it is bent must be considered to ensure the automation equipment does not interfere with the part during bending. Therefore each application should be reviewed prior to finalising the loading system design.

4.8 Miscellaneous manufacturing processes

The CNC controlled processes described so far are but a few of the processes that are available and used in engineering manufacture. It might be thought that automated

Fig. 4.37 *Contoured tooling set*

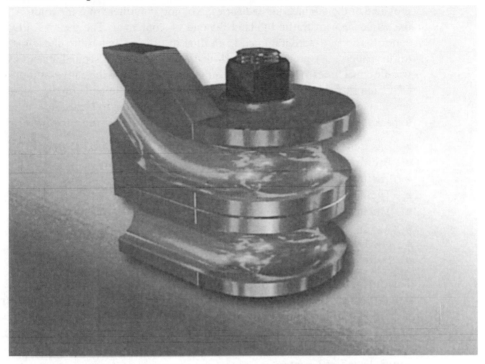

Fig. 4.38 *Automatic loading system*

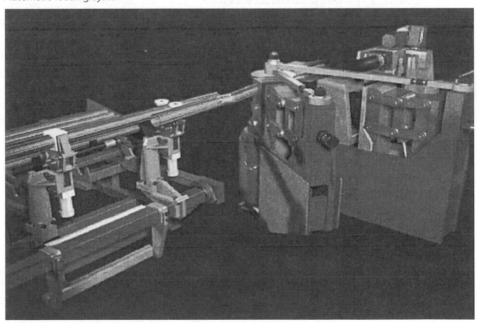

manufacture by CNC controlled machines and PLC controlled devices is the exclusive province of the engineering industries. Nothing is further from the truth. A PLC controls the sequencing of traffic lights at the end of your road. The use of a PLC enables the controller to be re-programmed from time to time to match the changing and evolving traffic patterns. A pre-programmed, dedicated controller sequences the events in your washing machine. When it comes to manufacturing, a wide range of industries employ CNC and PLC electronic data control devices to handle and convey solids, liquids and gases automatically; to form and process raw materials; to blend, mix and measure; and to package the end products automatically. Some typical, diverse industries are as follows:

- Electronic component and equipment manufacture and assembly.
- Brewing, distilling and the manufacture of soft drinks.
- The manufacture of petrochemicals.
- The manufacture of paints and emulsions.
- The manufacture and packaging of cosmetics and pharmaceuticals.
- The manufacture of fabrics and clothing.
- Food processing and packing.
- Wood machining for joinery products and furniture manufacture.

This list is only a sample of typical industries now using CAD/CAM in the manufacture of their products. There is not sufficient room in this chapter to deal with all aspects of these electronic data control applications but we will briefly consider the application of CNC to fabric weaving and wood machining. The uses of PLCs and industrial robots for process control and assembly, together with flexible manufacturing systems (FMSs) are discussed in Chapter 5.

4.8.1 *Fabric manufacture*

Joseph Marie Jacquard (1752–1834) invented a loom for weaving fabrics with intricate designs. The Jacquard loom was controlled by a chain of punched cards. These cards provided the information needed to weave a cloth with a particular pattern. By changing the chain of punched cards the pattern could be changed. By keeping and re-using the cards, subsequent batches of cloth could be woven with an identical pattern. Modern looms use CNC control in place of the chain of cards but the principle is the same. By extending the control to CAD/CAM the pattern design can be carried out on a computer and downloaded directly to the loom.

4.8.2 *Wood machining – electronic routing machine*

The SCM Routronic P-CU machine shown in Fig. 4.39 and manufactured by SCM of Italy is a typical example of a volume production CNC controlled routing machine capable of high volume production, yet flexible enough to manufacture small batches economically. Each parallel cutting head can be equipped with an automatic tool changer that has a ten-position tool magazine. This makes it possible to change component shapes very rapidly while maintaining the high productivity of a parallel head machine. This is but one type of machine manufactured by this company along with a wide range of interchangeable accessories to suit every wood machining application.

Fig. 4.39 *Routronic CNC wood machining*

This machine is controlled by ASTROCAD software that has the following features:

- Component drawing with simultaneous assignment of all machining operations.
- The CNC program is generated directly while the component is being drawn.
- It rapidly executes parametric drawings with respective CNC programs.
- It automatically executes pockets, recesses, cycles and preconfigured sub-programs.
- Drawings can be modified; the CNC program will be updated automatically with any drawing modifications.
- It simulates tool paths and work cycles making it possible to test a program before it is run.
- It calculates machining time.
- It calculates the cost of machined components.
- It will load part geometry in the self teach mode using masters or scale drawings with graphics tablets or digitizers.

ASTROCAD can be used with AUTOCAD to provide total integration between the design and manufacturing departments for off-line component design.

4.8.3 *Wood machining – flexible window processing production line*

In addition to individual CNC woodworking machines, as previously described, SCM also manufactures complete production line systems, for example the SCM 5S shown, diagrammatically, in Fig. 4.40.

Fig. 4.40 *SMC system 5S flexible window processing line*

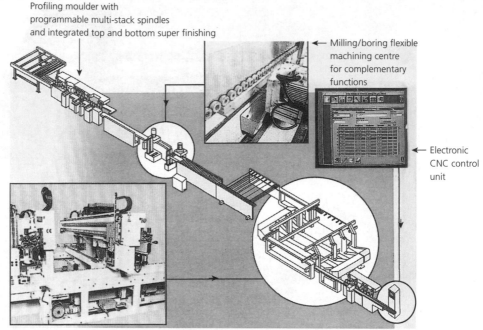

Profiling moulder with
programmable multi-stack spindles
and integrated top and bottom super finishing

← Milling/boring flexible
machining centre
for complementary
functions

← Electronic
CNC control
unit

Automatic double ended tenoner with workpiece
locking and feeding using CNC clamping

System 5S, the high technology window processing line, combines full operational flexibility with high production rates. It fully integrates the following machining operations: planing, tenoning, boring, profiling, execution of slots for hinges, automatic weather strip fitting and glazing bead mitre cut. This system has a high productive capacity for both large batches and single-piece working. The control system for supervising the production line (ASTROCAD) integrates three levels.

ASTROCAD is part of the machine. Once the drawing stage has been completed, the program is generated immediately without any further information because the software 'knows' the CNC codes for the machine, having already stored the machine configuration in memory.

Level 1

Level 1 is the machine level, and uses integrated and modular CNC/PLCs that are generally available in industry in order to optimise the performance of the individual machines, to increase reliability and to facilitate the procurement of spare parts.

Level 2

The line supervisor makes use of an industrial PC with user specific software that has been designed to offer XWINDOW/MOTIF with graphics, which help the operator to handle all functions:

- Centralised diagnostics to reduce intervention time when machine problems occur.
- The reporting of executed functions on the production line assisted by graphic monitoring.
- Pre-arrangement for remote service.
- Pre-arrangement for network link with network data processing systems.

Level 3

Level 3 provides supervision of all operational activities in the company. It can be completed with an appropriate software package, which handles the production orders and automatically elaborates on the jobbing list of the pieces to be produced, according to a customer purchase order. From the raw material fed into the machine, complete sets of machined components are automatically delivered from the 5S production line in the quantities and shapes required ready for assembly by the customer. The product mix can be pre-planned on a daily basis.

Conclusion

This is just a brief overview of the level of electronic data control that is applicable throughout manufacturing industry and is intended to show that process control is not the prerogative of the engineering and metalworking industries.

The authors would like to take this opportunity to thank the following persons and companies for their assistance in compiling this chapter and granting the use of their copyright material: Amada United Kingdom Ltd; VARI-FORM Inc., Ontario, Canada; Eagle Technologies, Canada; Jones and Shipman International, Leicester, UK; Derek J. Baker (Managing Director) of Centreless Tooling Co. Ltd, Tamworth, UK; SCM, Rimini, Italy.

ASSIGNMENTS

1. Compare and contrast the economics of manufacture of a sheet metal computer chassis by conventional pressing (stamping) with its manufacture by means of a turret punching machine.

2. Discuss the advantages and limitations of hydro-forming motor vehicle chassis components with their manufacture by conventional presswork. Pay particular attention to the economics of the processes.

3. Discuss the advantages and limitations of the application of CNC to precision-grinding processes.

4. Compare and contrast the economics of the centreless-grinding process compared with conventional cylindrical grinding for the mass of any cylindrical component with which you are familiar.

5. Describe and discuss the application of CNC to any **non-engineering** manufacturing process with which you are familiar, paying particular attention to the type of controller and software used, the advantages and limitations of the process, and whether or not adaptive control is appropriate.

5 The control of technology

When you have read this chapter you should be able to:

- discuss the control of technology;
- appreciate the use of programmable logic controllers;
- use a case study of a 'pick and place' unit to develop a PLC program;
- use the case study of a 'robot cell' to develop ideas on distributed control and the programming of a robot cell;
- analyse supervisory control and data acquisition;
- appreciate the use of teleservice monitoring equipment;
- appreciate the need for group technology;
- apply the principles of group technology to flexible manufacturing systems;
- apply the principles of three-dimensional coordinate measurement to measurement and quality control.

5.1 Introduction

Our daily lives are affected directly by automation, from the bread we eat to the cars we drive. We expect a high standard of living in the Western world but only want to pay the lowest price. The global economy is very competitive with contracts for the purchase of goods going to where the labour cost is lowest. That is why, in order to remain competitive, so many companies have had to adopt high levels of automation. For example:

- Bakeries where the ingredients are mixed automatically from giant hoppers. The dough is automatically weighed and cut to size to within critical limits and then dropped into tins, which slowly move along a controlled conveyor through infrared ovens and emerge as baked loaves ready for the customer.
- Pharmaceuticals are made in Class 100 clean room conditions having much stainless steel, as the environment and the equipment have to be biologically clean. The high levels of automation limit the risk of contamination from humans. Many critical phases such as the filling and heat-sealing of glass injection phials take place automatically on equipment that is housed behind sealed glass screens, where even higher levels of clean conditions exist.
- The production of petrochemicals and other hazardous chemicals is a highly controlled process because of the volatile nature of the materials. Also the level of

automation reduces the risk of humans coming into contact with the potentially dangerous and/or flammable chemicals and reduces the risk to the environment by human error.

- The manufacture of cars and the assembly of complex products can now be automated to a high degree. Again this reduces the risk of human error and the tedium of the operator having to repeat the same process over and over again. This frees the operator from drudgery and enables more focus to be placed on quality.

The following sections deal with how systems can be automated and controlled.

5.2 Automation and control

Many automated systems are used to control manufacturing processes, for example the production and packaging of chemicals and pharmaceuticals and the control and monitoring of assembly systems. No matter what the application, they all have the same building blocks of control.

- *Inputs* – signals from sensors, both digital and analogue.
- *Outputs* – signals to switch solenoid valves or turn on motors.
- *Feedback* – monitoring of the process in order to compare the desired result with the actual result, for example a safety check.

The complexity of modern automated systems means that individual items of equipment all have their own onboard, dedicated computer, which communicates to other items of equipment, for example robots communicating with machine tools to see if they are ready to be loaded or unloaded. The individual systems can then be regarded as *cells of automation*. If they cannot communicate outside of the cell to the larger system then they are known as *islands of automation*.

The main area of impact that e-manufacture has had on many sectors of industry is its ability to network together diverse systems and machines. For example *distributed control* can be used within a factory where intelligent machines can communicate with one another. It is the same as the computer integrated manufacture (CIM) pyramid that is shown in Fig. 5.1. This has the human operator at the top level with ultimate control over the computerised *supervisory control and data acquisition* (SCADA), which, in turn, is in command over the *cell controller* (PLC), right down to individual machines and sensors. Remember a PLC is a programmable logic controller.

Fig. 5.1 *Computer integrated manufacture (CIM) pyramid of command*

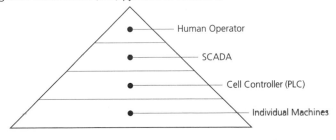

The World Wide Web (WWW) is currently linked to SCADA systems mainly on a reporting basis. It would not be wise to control equipment over the Web, although the advent of a Web programming language called JAVA and intelligent equipment (Blue Tooth) may see this happen on a commercial basis in the near future. However, in this application, the Internet has had limited use due to security problems and data protection, that is, controlling dangerous equipment via the Internet could be hazardous. Internet facilities, if used, are limited to read-only. At present, many corporations use ISDN lines to communicate to their control systems. If the security of the Internet could be guaranteed and interaction was possible, the Internet would be a consideration.

5.3 Programmable logic controllers

The latest range of PLCs will work within a communication network and can be programmed off-line very easily. They use standard serial or Modbus communication connections and protocols. This enables individual items of equipment to communicate with other items of equipment either via their own internal PLCs or via external devices. Thus PLCs are able to communicate with each other via a computer network. Master computers can form another level of control to monitor the whole process. Many PLCs now use off-line programming techniques.

The examples in this chapter use FLEX, which is a software package for programming miniature PLCs, such as the Colter FMT100 and FMT200 that are widely used for manufacturing process controls in industries ranging from brewing to petrochemicals. Such controllers are also suitable for pick and place robots used for dedicated assembly. The instruction language is text based and can be easily learned by anyone, especially those with a knowledge of computer languages – BASIC, C, Pascal, etc.

The FMT100 and the FMT200 controllers are based upon PLCs that originally were developed by the manufacturer Saab to control car and lorry production and enabled the next generation of intelligent controllers to communicate with each other and to a central computer. Projects can be written in a combination of *ladder* and *instruction languages*. Instruction language modules can be executed on a *scan*, *time* or *interrupt* basis. All the instruction language modules in a project execute (work) simultaneously allowing parallel processes to operate independently. The language uses keyword statements to control execution, such as:

- IF . . . THEN . . . ELSE
- WHILE
- DO . . . UNTIL
- FOR . . . NEXT
- REPEAT

5.3.1 *FLEX*

This software is very flexible and has a number of useful features:

- Graphical ladder programming (ladder logic).
- Text-based high-level language programming (instruction set).

- Source level debugging.
- Syntax highlighting in both programming languages.

Let us now examine two case studies to find out how automation and control can be achieved.

5.4 Case study 1 – pick and place cell

First we need to know how to write a program using the FLEX instruction set language. The program must allow a user to specify any number of components (up to a maximum of eight in this example) that are to be loaded onto a conveyor belt using a pneumatic pick and place unit. The unit can be seen in Fig. 5.2. For simplicity, wooden blocks are being used for the components in this example.

5.4.1 *Initial analysis*

The inputs (I) are at the top of the FMC100 micro PLC shown in Fig. 5.3(a) and are numbered (i.e. I0, I1, etc.) and the outputs (Q) were also in order (i.e. Q0, Q1, etc.) at the bottom. These lights indicated the specific conditions for a particular action, i.e. to understand which of the signals had to be true or not before the next operation would execute. These signals control different parts of the operation, i.e. arm extend or retract, and were identified on the unit shown in Fig. 5.2 as I2, I3, Q2 and Q1.

A range of optical, limit and proximity sensors on the pick and place unit provide the input (I) signals. The output (Q) signals are sent to a bank of pneumatic solenoid

Fig. 5.2 *Pick and place robot*

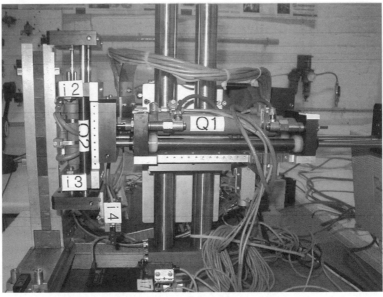

Fig. 5.3 *Pick and place robot control: (a) control unit; (b) schematic diagram*

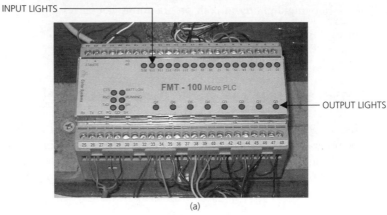

(a)

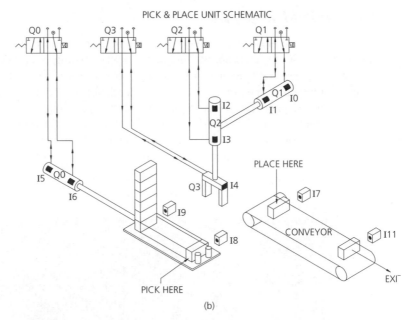

(b)

valifes in response to the inputs, with each one controlling a different pneumatic cylinder on the unit. These are shown in Fig. 5.3(b), which is a schematic diagram of the system.

5.4.2 *Sequence of events*

The sequence of events required to operate the pick and place unit is now described. It is also illustrated as a flowchart in Fig. 5.4.

1. *Piston Q0 moves forward pushing a block up to a stop in readiness for being picked up.* This will only happen if the following conditions apply:

Fig. 5.4 *Flowchart for pick and place operation*

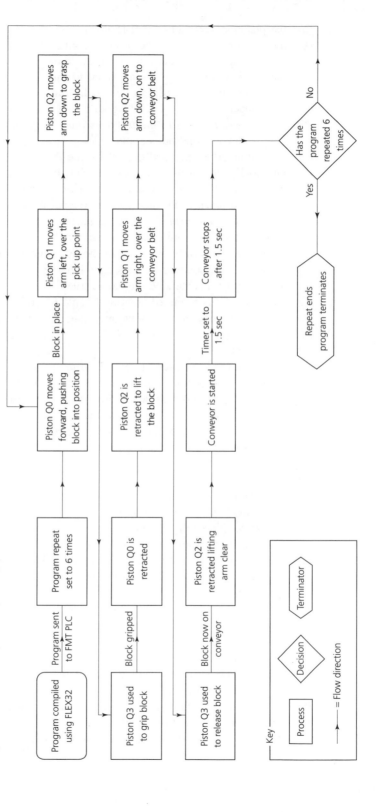

(a) the piston is retracted (signal I5 true);

(b) there is a block ready to be pushed (signal I9 true);

(c) the pick up area is clear (signal I8 not true);

(d) the arm is in the left position (signal I0 true);

(e) the arm is in the up position (signal I2 true);

(f) the gripper has nothing in it (signal I4 not true).

2. *Piston Q1 moves the arm to the left over the pickup point.* This will only happen when the following conditions apply:

(a) the piston Q0 is extended (signal I6 true);

(b) the block is in position for picking (signal I8 true);

(c) the piston Q2 is in its retracted up position (signal I2 true);

(d) there is no block in the storage area (signal I9 not true).

3. *Piston Q2 moves the arm down in readiness to grasp a block.* This will only happen when the following conditions apply:

(a) the piston Q1 is fully extended to the left (signal I1 true);

(b) the block is in position for picking (signal I8 true);

(c) the piston Q0 is extended (signal I6 true);

(d) the arm is in its retracted position (signal I2 true);

(e) there is no block in the storage area (signal I9 not true).

4. *Piston Q3 is used to grip the block.* This will only happen when the following conditions apply:

(a) the piston Q0 is extended (signal I6 true);

(b) the block is in position for picking (signal I5 true);

(c) the piston Q2 is fully extended down (signal I3 true);

(d) the piston Q1 is fully extended to the left (signal I3 true);

(e) there is no block in the storage area (signal I9 not true);

(f) the gripper has nothing in it (signal I4 not true).

5. *Loading piston Q0 is retracted.* This will only happen when the following conditions apply:

(a) the piston Q0 is extended (signal I6 true);

(b) the block is in position for picking (signal I8 true);

(c) the piston Q2 is fully extended down (signal I3 true);

(d) the gripper has grasped the block (signal I4 true);

(e) there is no block in the storage area (signal I9 not true).

6. *Piston Q2 is retracted to lift the block.* This will only happen when the following conditions apply:

(a) the piston Q0 is fully retracted (signal I5 true);

(b) there is a new block ready to be pushed (signal I9 true);

(c) the pickup area is clear (signal I8 not true);

(d) the piston Q2 is fully extended down (signal I3 true);

(e) the piston Q1 is fully extended to the left (signal I1 true);

(f) the gripper has grasped the block (signal I4 true).

7. *Piston Q1 is retracted to move the block over the conveyor.* This will only happen when the following conditions apply:

(a) the piston Q0 is fully retracted (signal I5 true);

(b) there is a new block ready to be pushed (signal I9 true);

(c) the piston Q2 is in its retracted up position (signal I2 true);

(d) the piston Q1 is fully extended to the left (signal I1 true);

(e) the gripper has grasped the block (signal I4 true);

(f) the pickup area is clear (signal I8 not true).

8. *Piston Q2 moves the arm down to lower the block onto the conveyor.* This will only happen when the following conditions apply:

(a) the piston Q0 is fully retracted (signal I5 true);

(b) there is a new block ready to be pushed (signal I9 true);

(c) the piston Q2 is in its retracted up position (signal I2 true);

(d) the piston Q1 is fully retracted to the right (signal I0 true);

(e) the gripper has grasped the block (signal I4 true);

(f) the pickup area is clear (signal I8 not true);

(g) the sensor at the lowering point is clear (signal I7 not true).

9. *Piston Q3 releases the block.* This will only happen when the following conditions apply:

(a) the piston Q0 is fully retracted (signal I5 true);

(b) there is a new block ready to be pushed (signal I9 true);

(c) the piston Q2 is fully extended down (signal I3 true);

(d) the piston Q1 is fully retracted to the right (signal I0 true);

(e) the gripper has grasped the block (signal I4 true);

(f) there is a block in position on the conveyor (signal I7 true);

(g) the pickup area is clear (signal I8 not true).

10. *Piston Q2 retracts, lifting the arm clear.* This will only happen when the following conditions apply:

(a) the piston Q0 is fully retracted (signal I5 true);

(b) there is a new block ready to be pushed (signal I9 true);

(c) the piston Q2 is fully extended down (signal I3 true);

(d) the piston Q1 is fully retracted to the right (signal I0 true);

(e) there is a block in position on the conveyor (signal I7 true);

(f) the pickup area is clear (signal I5 not true);

(g) the gripper has nothing in it (signal I4 not true).

11. *The conveyor is started.* This will only happen when the following conditions apply:

(a) the piston Q0 is fully retracted (signal I8 true);

(b) there is a new block ready to be pushed (signal I9 true);

(c) the piston Q2 is in its retracted up position (signal I2 true);

(d) the piston Q1 is fully retracted to the right (signal I0 true);

(e) there is a block in position on the conveyor (signal I7 true);

(f) the pickup area is clear (signal I8 not true);

(g) the gripper has nothing in it (signal I4 not true);

(h) the sensor at the conveyor is clear (signal I11 not true).

12. *The Conveyor operates for 1.5 seconds in order to allow the block to be moved clear.* This happens when the conveyor belt starts moving and the timer is set for 1.5 sec, then the belt stops.

5.4.3 *Ladder logic*

The principle of *ladder logic* programming is very simple, before a command executes a number of states have to be satisfied. The ladder program derived from Fig. 5.4 is shown in Fig. 5.5(a). The ladder program is developed graphically within the FLEX32 program, where the symbols are dragged and dropped into their appropriate sequence on the ladder diagram. Figure 5.5(b) shows the first line (or rung of the ladder) and is interpreted as:

- IF I5 AND I9 AND I2 are TRUE
- THEN I8 AND I4 are NOT TRUE
- THEN TURN-ON OUTPUT Q0

Unless these states are satisfied then Q0 will not be turned on. There is also a *latch* called F000, this is a reset latch. Its purpose is to make sure that once Q0 is switched

Fig. 5.5 *Ladder logic: (a) logic derived from the flow chart; (b) first rung*

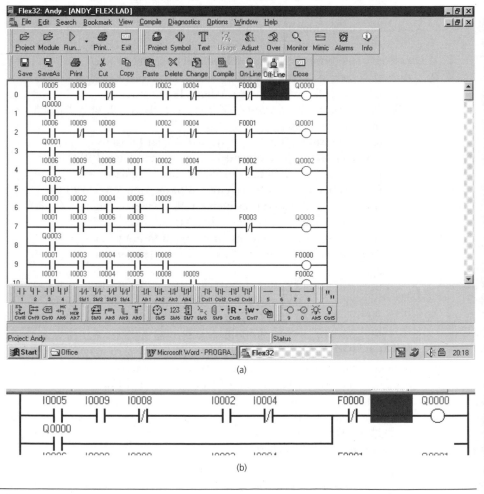

(a)

(b)

on it is not turned off until instructed, hence the Q0 on the second line, which appears to bypass all the inputs, i.e. when on it will ignore the input states. The reason is that this ensures safe operation and it will be explained later in the chapter.

5.4.4 *Instruction set*

Each block of code will tell you which function it will control. Since it is an instruction set it is in a lowercase font as FLEX generates the commands automatically in this format. An example of part of the control program to move the arm down as written in instruction set code is shown as follows:

*Move Arm (Piston q2) . . . this is a remark
wait_for *i1* **and** *i8* **and** *i6* **and** *i2*
wait_not *i9*
turn_on *q2*

Here you can see the code, which moves the arm down to grasp the block by turning on signal q2, the label at the top tells you what this specific section of code will do. As mentioned it is simple to understand and this section interprets as, turn q2 on when i1, i8, i6 AND i2 are TRUE AND i9 is NOT TRUE. The remainder of the program is completed in the same way, which allows the programmer to follow the code operation by operation, as shown in Fig. 5.6.

Text strings

It is useful to print operator messages to the computer screen, which is connected to the FLEX PLC by a serial lead. For example this enables the operator to receive safety commands or receive machine utilisation data. Text strings can be up to 75 characters long and any ASCII printable character can be entered. Each string is given a *Text Number*, which is pre-set down the left-hand side of the sheet. For example, the text string called within the program by using the TEXT command:

wait for *P1free*
text (tx23, 1)

Here text string 23 will be transmitted from serial port 1 on the PLC to the computer. By using this command, operator messages can be produced. This provides the feedback and interactivity that is required for both safety and production information.

Interaction

To alter the number of blocks to be moved, the program can be altered by changing the number of times the program or instruction is repeated, for example the following program segment:

*Enter Number Of Times To Repeat Program Here.
repeat 6

Remember that the asterisk indicates that the statement is a program 'remark'.

Fig. 5.6 *Control program written instruction set language*

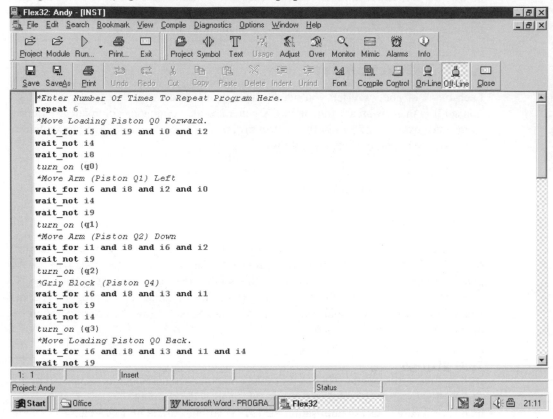

This command will instruct the program to loop 6 times. That is, it will operate the unit for 6 pick and place operations thus moving six blocks. By altering this value and saving the program the user can change the number of blocks to be moved. A more sophisticated way of making the program interactive is to set the repeat value as a variable which can be inputted via the keyboard.

5.4.5 *Safety issues*

Safety issues need to be addressed, with equal importance no matter whether the processes are automated or manually operated. When the program runs, the first step is for a block to be pushed into place against location stops as shown in Fig. 5.7(a) by the use of a pneumatically-actuated piston. If the piston is released immediately, the block will bounce back from the stops. If the arm then lowers to pick the block up it is possible that it may have bounced out of reach or may interfere with the lowering of the arm, causing damage. In a factory this could be very dangerous so, to combat this problem, we have to ensure that the piston remains extended thus making sure the block does not

Fig. 5.7 *Pick and place robot in action: (a) block located safely against location stops; (b) arm fully retracted*

Block

Stop

(a)

(b)

move until it is has been grasped securely. It is important to ensure that the lifting arm will be fully retracted before the block is moved over the conveyor belt as shown in Fig. 5.7(b). This is necessary to eliminate the risk of collision during movement. If we lift the block while moving it over the conveyor, there is a risk of collision; by ensuring that the arm is retracted we eliminate this risk.

The next area to consider is the loading of a block onto the conveyor belt. If the belt is moving when the block is lowered, it may bounce or be dragged out of the gripper; again this could be dangerous, as the part may not sit correctly on the belt. This is not a problem if the part is not being used again, but if it needs to be picked up later it may cause delays. In order to prevent this happening the code must ensure that the belt does not start moving until the block has been placed in position, the gripper is then released and the arm is moved clear. The belt is then started to ensure that the block is moved to a safe position before the next block is loaded. After a short delay the belt is stopped to allow the safe loading of the next block.

Another important issue is to ensure that the block is not released until it has been firmly placed on the belt. The reason for this is that if the block is released early then there is a risk of it dropping and bouncing out of position on the belt. In industry the dropping of heavy components could be detrimental to machinery and dangerous to human life. If the part is being lowered into liquid, such as in coating operations, it could cause dangerous splashing of the liquid. Although this example involves only a small-scale pick and place unit, it forces you to consider the implications of what may happen and then provide adequate safety considerations within your code. The complete program is shown in Table 5.1. It is important to become aware of the safety issues involved in programming before moving onto larger industrial scale equipment as in case study 2.

Table 5.1 *The program for the pick and place robot*

*Enter Number of Times To Repeat
Program Here.*
repeat 6
Move Loading Piston Q0 Forward.
wait_for i5 **and** i9 **and** i0 **and** i2
wait_not i4
wait_not i8
turn_on (q0)
Move Arm (Piston Q1) Left
wait_for i6 **and** i8 **and** i2 **and** i0
wait_not i4
wait_not i9
turn_on (q1)
Move Arm (Piston Q2) Down
wait_for i1 **and** i8 **and** i6 **and** i2
wait_not i9
turn_on (q2)

Table 5.1 *(cont'd)*

Grip Block (Piston Q3)
wait_for i6 **and** i8 **and** i3 **and** i1
wait_not i9
wait_not i4
turn_on (q3)
Move Loading Piston Q0 Back.
wait_for i6 **and** i8 **and** i3 **and** i1 **and** i4
wait_not i9
turn_off (q0)
Move Arm up (Piston Q2) Up
wait_for i5 **and** i9 **and** i8 **and** i3 **and** i1
and i4
turn_off (q2)
Move Arm (Piston Q1) Right
wait_for i5 **and** i9 **and** i2 **and** i1 **and** i4
wait_not i8
turn_off (q1)
Move Arm (Piston Q2) Down
wait_for i5 **and** i9 **and** i2 **and** i0 **and** i4
wait_not i8
wait_not i7
turn_on (q2)
Release Grip On Block (Piston Q3)
wait_for i5 **and** i9 **and** i3 **and** i0 **and** i4 **and** i7
wait_not i8
turn_off (q3)
Move Arm up (Piston Q2) Up
wait_for i5 **and** i9 **and** i3 **and** i0 **and** i7
wait_not i8
wait_not i4
turn_off (q2)
Start Conveyor Belt
wait_for i5 **and** i9 **and** i2 **and** i0 **and** i7
wait_not i8
wait_not i4
wait_not i11
turn_off (q5)
Set Delay For Conveyor (t0) to 1.5 seconds
wait_for (q5)
on_delay (t0, 0, 0, 1, 50)
Wait For Set Delay (t0) Time Then Stop Conveyor
wait_for t0
turn_off (q5)
End The Repeat Set At The Start Of The Program
end_repeat

5.5 Case study 2 – the flexible assembly cell

This section examines how a flexible assembly cell may be developed using an industrial robot and a palletised conveyor system with distributed control

5.5.1 *Distributed control*

The example used is a robot that loads a pallet at station 1 with bolts and then unloads the pallet at station 2, as shown in Fig. 5.8(a) and the details are shown in Figs 5.8(b) to 5.8(d) inclusive.

Fig. 5.8 *Robot cell: (a) plan view of PUMA track layout.*

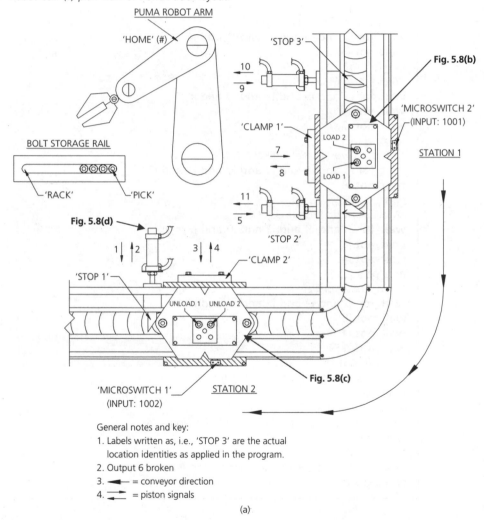

General notes and key:
1. Labels written as, i.e., 'STOP 3' are the actual location identities as applied in the program.
2. Output 6 broken
3. ◄— = conveyor direction
4. ⇄ = piston signals

(a)

Fig. 5.8 *(b) pallet at loading station; (c) pallet at un-loading station*

Stop

Clamp 1

FMT PLC

(b)

Rack Point

Pick Point

Clamp 2

Microswitch

(c)

At the top of the hierarchical chain is the operator who observes the supervisory, control and data acquisition display that is a graphical representation of the entire system: the SCADA system. This system is responsible for:

- Interaction with the operator.
- Reporting alarms and faults.
- Monitoring progress – i.e. reporting trends.
- Enabling the operator to control certain aspects of the system if required.

Fig. 5.8 *(d) un-loading station waiting for pallet to arrive*

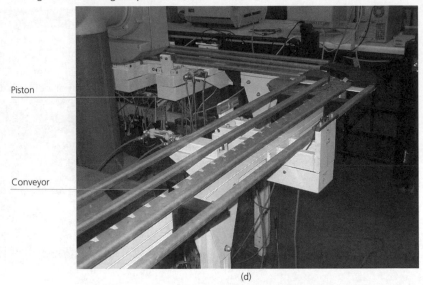

(d)

The master PLC communicates with the SCADA computer via a serial communications link. The state of the inputs and outputs is communicated to the SCADA system that represents, graphically, the device being controlled – i.e. the robot. The master PLC is responsible for monitoring all inputs – i.e. light sensors – and activating the outputs accordingly. The program flow chart is shown in Fig. 5.9.

5.5.2 *Robot programming*

The program shown in Table 5.2 was written and tested offline using *VAL2* robot programming language to accomplish the following sequence:

(a) Sense and trap the pallet at station 1 (FMT 200 PLC).
(b) Load 3 bolts in the wooden block (program TASK 1) then release the pallet.
(c) Sense and trap the pallet at station 2 (FMT 200 PLC).
(d) Unload the bolts back into the rack then release the pallet (program TASK 2).
(e) Repeat the process for 2 bolts and then 1 bolt (program MAIN).
(f) Use arrays, text prompts and sub-programs for an interactive and concise program.

5.5.3 *Interaction between the PLC and the robot arm*

There are two outputs (1 and 2) on the robot connected to two inputs (I20 and I21) on the PLC conveyor system, and two outputs (Q11 and Q12) on the PLC controlled conveyer system connected to two inputs (1001 and 1002) on the robot as illustrated in Fig. 5.9. Both these devices can set their outputs to logical one (+24 volts) or logical zero (0 volts). The outputs of both the robot and the PLC controlled conveyor system are under software control and communicate with each other, thus when one device is operating the other device is waiting for a signal to reverse the situation. Only one of the two devices is in control of the cell at any one time. In this way the control of the cell is passed safely between the two programmed devices. The output signal need only be present for a short time.

The next command is 'SIGNAL −3' (this means 'signal 3 not on'), which sets output three to logical zero (0 volts). This switches off the conveyor belt. A sub-program called 'task 1' is now used within the robot program. This program moves the robot arm so that it takes the bolts from the rack and places them into the holes in the block on the pallet.

5.5.4 *Basic working program*

The basic program and sub-programs can be seen in full in Table 5.2. The master program is called 'main' and starts with the keyword 'RESET' this resets all the output signals from the robot to a logical zero (0 volts). This sets the bolt assembly cell to a known state. **It is an important safety rule to start a new program with this command as the robot could be left in a dangerous configuration by the previous program.** Safety is the primary consideration when writing any program that controls any moving device. Operator messages are achieved in this program by using the command 'TYPE'. This command displays text on the computer screen. The command 'PROMPT' requests any input from the operator.

Note! All locations are described as upper case names, i.e. PICK, RACK, LOAO(X) and UNLOAD(Y). All program names are in lower case i.e. main, load and unload.

5.6 Comparison of the two case studies

The two case studies use different technologies to solve similar problems. In order to select the optimum technology the two will be compared in relation to cost, accuracy, drive and feedback systems and applications.

5.6.1 *Comparison of cost*

The cost of a typical industrial robot depends upon a number of factors:

- Complexity and the degrees of freedom. This is a measure of the number of moving joints.
- The size and maximum load it has to move – robots can vary in size from the very small to the extremely large.
- The amount of expected backlash and positioning error.
- Reliability and speed of movement.

The cost of a PLC can be as little as £100 and with pneumatic rams and conveyors the total cost of a pick and place unit can be as little as a tenth of the cost of a robot. The approximate cost of the pick and place unit in this case is £3500 whereas the cost of a comparable industrial robot is probably nearer £70 000.

Fig. 5.9 *Flowchart for PUMA. TXT VAL II program*

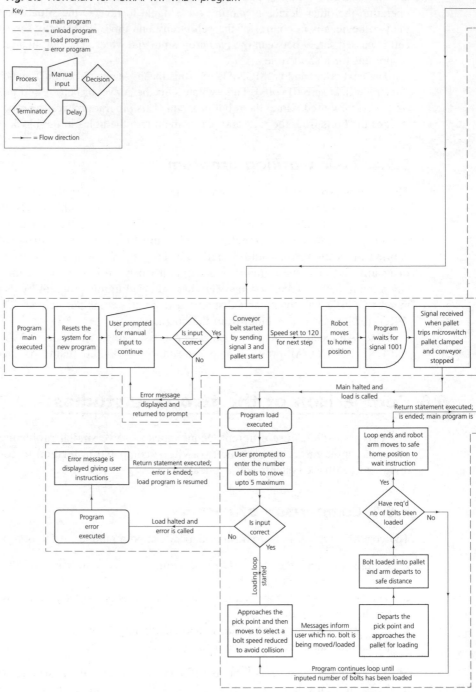

Key

- – – – – = main program
- – – – – = unload program
- – – – – = load program
- – – – – = error program

Process

Manual input

Decision

Terminator

Delay

→ = Flow direction

Program main executed → Resets the system for new program → User prompted for manual input to continue → Is input correct → Yes → Conveyor belt started by sending signal 3 and pallet starts → Speed set to 120 for next step → Robot moves to home position → Program waits for signal 1001 → Signal received when pallet trips microswitch pallet clamped and conveyor stopped

No → Error message displayed and returned to prompt

Main halted and load is called

Return statement executed; is ended; main program is

Program load executed

Error message is displayed giving user instructions ← Return statement executed; error is ended; load program is resumed → User prompted to enter the number of bolts to move upto 5 maximum

Loop ends and robot arm moves to safe home position to wait instruction

Yes

Program error executed ← Load halted and error is called → Is input correct

No

Yes → Loading loop started

Have req'd no of bolts been loaded → No

Bolt loaded into pallet and arm departs to safe distance

Approaches the pick point and then moves to select a bolt speed reduced to avoid collision → Messages inform user which no. bolt is being moved/loaded → Departs the pick point and approaches the pallet for loading

Program continues loop until inputed number of bolts has been loaded

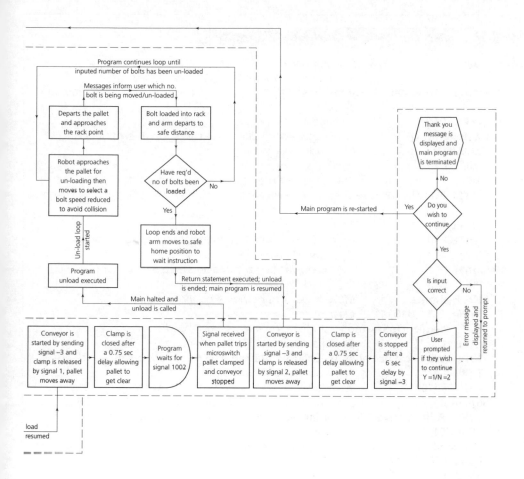

Program continues loop until inputed number of bolts has been un-loaded

Messages inform user which no. bolt is being moved/un-loaded

| Departs the pallet and approaches the rack point | Bolt loaded into rack and arm departs to safe distance |

Robot approaches the pallet for un-loading then moves to select a bolt speed reduced to avoid collision

Have req'd no of bolts been loaded — No

Yes

Loop ends and robot arm moves to safe home position to wait instruction

Un-load loop started

Program unload executed

Return statement executed; unload is ended; main program is resumed

Main halted and unload is called

Thank you message is displayed and main program is terminated

No

Main program is re-started — Yes — Do you wish to continue

Yes

Is input correct — No

Conveyor is started by sending signal –3 and clamp is released by signal 1, pallet moves away

Clamp is closed after a 0.75 sec delay allowing pallet to get clear

Program waits for signal 1002

Signal received when pallet trips microswitch pallet clamped and conveyor stopped

Conveyor is started by sending signal –3 and clamp is released by signal 2, pallet moves away

Clamp is closed after a 0.75 sec delay allowing pallet to get clear

Conveyor is stopped after a 6 sec delay by signal –3

User prompted if they wish to continue Y =1/N =2

Error message displayed and returned to prompt

load resumed

Table 5.2 *The PUMA cell program including annotation*

```
.program main                              ;Program name
       RESET                               ;Resets from previous user
    20 Prompt "To Initiate Program Press 1:", R   ;Prompt for user to initiate program by
                                           pressing 1
       TYPE /C1,""                         ;Carriage return
       CASE INT(R) OF                      ;Case statement to check validity of input, error handling
       VALUE 1:                            ;Check for input value of 1
         GOTO 10                           ;If value is 1 then goto 10
       ANY                                 ;Any statement used for any other input value
         TYPE "ERROR!: Please Press 1 To Proceed"   ;Error message displayed on screen,
                                           invalid input
         TYPE /C1, ""                      ;Carriage return
         GOTO 20                           ;Goto 20 to prompt user to initiate the program again
       END                                 ; End case statement
    10 Signal 3                            ;Start conveyor belt by sending signal 3
       SPEED 120                           ;Set speed to 120 for the next step only
       MOVE #Home                          ;Move Robot to safe home position
       TYPE /C23, /U18, ""                 ;Clear screen
       TYPE "Sequence Started"             ;Information message displayed
       TYPE /C1, ""                        ;Carriage return
       TYPE "Waiting For Pallet To Load At Station 1"   ;Information message displayed
       TYPE /C1, ""                        ;Carriage return
       WAIT Sig(1001)                      ;Wait for sig 1000 from PLC before executing next step
       Signal -3                           ;Stop the conveyor belt by sending signal -3
       TYPE "Pallet Arrived At Loading Station"   ;Information message displayed
       TYPE /C1, ""                        ;Carriage return
       CALL load                           ;Call load program and execute
       Signal 3                            ;Start Conveyor by sending signal 3
       DELAY 0.5                           ;Delay program for 0.5 sec
       Signal 1                            ;Open clamp at load station by sending signal 1
       TYPE "Pallet Leaving Loading Station"   ;Information message displayed
       TYPE /C1, ""                        ;Carriage return
       DELAY 0.75                          ;Delay program for 0.75 sec
       Signal -1                           ;Close clamp at load station by sending signal -1
       TYPE "Waiting For Pallet To Un-Load At Station 2"   ;Information message
                                           displayed
       TYPE /C1, ""                        ;Carriage return
       WAIT Sig(1002)                      ;Wait for Signal 1002 from PLC before executing next step
       Signal -3                           ;Stop the conveyor belt by sending signal -3
       TYPE "Pallet Arrived At Un-Loading Station"   ;Information message displayed
       TYPE /C1, ""                        ;Carriage return
       CALL unload                         ;Call unload prog and execute
       Signal 3                            ;Start Conveyor belt by sending signal 3
       DELAY 0.5                           ;Delay program for 0.5 sec
       Signal 2                            ;Open clamp at unloading station by sending signal 2
       TYPE "Pallet Leaving Un-Loading Station"   ;Information message displayed
       TYPE /C1, ""                        ;Carriage return
       DELAY 0.75                          ;Delay program for 0.75 sec
```

Table 5.2 *(cont'd)*

```
          Signal -2                    ;Close clamp at unloading station by sending signal -2
          DELAY 6                      ;Delay program for 6 sec to allow pallet to clear
          Signal -3                    ;Stop the conveyor by sending signal -3
       60 Prompt "Do You Wish To Continue Y = 1/N = 2:", T   ;Prompt user if they wish to
                                        continue
          TYPE /C1, ""                 ;Carriage return
          CASE INT(T) OF               ;Case statement to check for valid input
          VALUE 1:                     ;Check for input value of 1
            GOTO 10                    ;If value is 1 then goto 10 and start program again
          VALUE 2:                     ;Check for input value of 2
            TYPE "THANKYOU FOR USING THEIR PROGRAM"   ;If value is 2, display message
            TYPE /C1, ""               ;Carriage return
            GOTO 50                    ;Goto to 50 to end program
          ANY                          ;Any statement used for any other input value
            TYPE "ERROR!: Please Press 1 For Yes Or 2 For No"   ;Error message displayed
                                        due to invalid input
            TYPE /C1, ""               ;Carriage return
            GOTO 60                    ;Goto 60 to re-prompt the user for an input
       50 END                          ;End case statement
.END                                   ;End program and terminate

.Program load                          ;Program name
       30 PROMPT "Select Number Of Bolts To Move & Press Return:", P   ;Prompt user for
                                        number of bolts to move
          TYPE /C1, ""                 ;Carriage return
          CASE INT(P) OF               ;Case statement to check for valid input
          VALUE 1,2,3,4,5:             ;Check if value is 1,2,3,4, or 5
            GOTO 40                    ;If value correct then goto 40 to continue
          ANY                          ;Any statement used for any other input value
            CALL Error                 ;Call error program
            GOTO 30                    ;Goto 30 to re-prompt for input value
          END                          ;End case statement
       40 TYPE "Thankyou, You Have Elected To Move",P, "Bolt(s), Please Wait"
                                        ;Information message displayed
          TYPE /C1, ""                 ;Carriage return
          For Q = 1 to P               ;Assign variable of Q to number of bolts to be moved P to
                                        set load loop
            APPRO PICK, 100            ;Approach the pre-programmed pick point at a height of 100
            OPENI                      ;Open gripper
            SPEED 20                   ;Set speed to 20, this step only, slow to avoid collision
            MOVES PICK                 ;Move to the pre-programmed pick point
            DELAY 0.75                 ;Delay program for 0.75 sec
            CLOSEI                     ;Close gripper to hold bolt
            TYPE "Bolt Number",Q, "Picked From Rack"   ;Tell user which number bolt is
                                        being moved
            TYPE /C1, ""               ;Carriage return
            DELAY 0.75                 ;Delay program for 0.75 sec
            DEPART 100                 ;Depart pick point to a height of 100
            APPRO LOAD[Q], 100         ;Approach load point of bolt number Q at a height of 100
            SPEED 20                   ;Set speed to 20, this step only, slow to avoid collision
            MOVES LOAD[Q]              ;Move to load point with bolt number Q
```

Table 5.2 *(cont'd)*

```
              TYPE "Loading Bolt Number", Q    ;Inform user which number bolt is being loaded
              TYPE /C1, ""            ;Carriage return
              DELAY 0.75             ;Delay program for 0.75 sec
              OPENI                  ;Open gripper to release the bolt
              DELAY 0.75             ;Delay program for 0.75 sec
              DEPART 100             ;Depart the load point at a height of 100
              CLOSEI                 ;Close gripped
            END                      ;End loop when Q number of bolts have been loaded
            SPEED 120                ;Set speed to 120 for next step only
            MOVE #Home               ;Move robot to safe home position
            RETURN                   ;Return to the main program
  .END                               ;End load program


.Program unload                      ;Program name
            For S = 1 to P           ;Assign variable of S to number of bolts to be moved P to
                                       set un-load loop
              APPRO UNLOAD[S], 100   ;Approach un-load point of bolt number S at a height of
                                       100
              OPENI                  ;Open gripper
              SPEED 20               ;Set speed to 20, this step only, slow to avoid collision
              MOVES UNLOAD[S]        ;Move to un-load point of bolt number S at a height of 100
               TYPE "Un-Loading Bolt Number", S   ;Inform user which bolt is being moved
               TYPE /C1, ""          ;Carriage return
              DELAY 0.75             ;Delay program 0.75 sec
              CLOSEI                 ;Close gripper to hold bolt
              DELAY 0.75             ;Delay program 0.75 sec
              DEPART 100             ;Depart load position to a height of 100
              APPRO RACK, 100        ;Approach the rack point at a height of 100
              SPEED 20               ;Set speed to 20, this step only, slow to avoid collision
              MOVES RACK             ;Move to the rack point at a height of 100
              DELAY 0.75             ;Delay program for 0.75 sec
              OPENI                  ;Open gripper to release the bolt
               TYPE "Bolt Number", S, "Placed In Rack"   ;Inform user which bolt has been
                                       placed in rack
               TYPE /C1, ""          ;Carriage return
              DELAY 0.75             ;Delay program 0.75 sec
              DEPART 100             ;Depart the rack point to a height of 100
              CLOSEI                 ;Close gripper
            END                      ;End loop if S number of bolts has been unloaded
            SPEED 120                ;Set speed tp 120 for next step only
            MOVE #Home               ;Move to safe home position
            RETURN                   ;Return to main program
  .END                               ;End unload program


.Program ERROR                       ;Program name
            TYPE "ERROR! Please Select A Number Of Bolts Between 1 And 5"   ;Display
                                       error message
            TYPE /C1, ""             ;Carriage return
            RETURN                   ;Return to main program
  .END                               ;End Error program
```

5.6.2 *Comparison of accuracy*

A robot arm has relatively good accuracy due to its drives and feedback systems, which will be discussed later. Since the arm relies on many joints – degrees of freedom – to move, any small amounts of 'play' in each of the joints becomes cumulative. Thus feedback systems, for example optical encoders, have to be good to ensure that the arm can move with pinpoint accuracy.

Conversely, a pick and place unit is extremely accurate, as it has fewer degrees of freedom. This is mainly because the parts of a pick and place unit tend to be either open or closed so that they are always within exact parameters. For example a pneumatic ram will always finish in the same place every time thus ensuring accurate repeatability. This makes it perfect for repetitive tasks.

5.6.3 *Comparison of drive and feedback systems*

The robot drive axis of movement can either be linear or rotary (revolute) and in some cases a combination of the two is used. These movements are achieved by using various mechanical devices to convert the rotation of, say, the electric motor into precisely controlled linear arm movement. Some common types of mechanism are:

- leadscrews and nuts;
- recirculating ball screws;
- chain and linkage drives;
- gear drives;
- belt drives;
- harmonic drives.

A robot is a complex device with many degrees of freedom and therefore requires a complex control with a closed-loop system, which employs feedback to ensure that the end effector is positioned accurately. Algorithms within the computer control determine both the angular and Cartesian positions within each joint. The end effector position is defined as a Cartesian coordinate and the roll, pitch and yaw of the wrist. The algorithm uses inverse kinematics to calculate each joint orientation.

A dedicated device, such as a pick and place unit, does not need to have such accurate control and feedback. Pick and place robots employ open-loop systems with no feedback, for example when a pneumatic ram is activated because the robot element being moved is controlled positionally by positive stops and proximity switches. Because of its comparative simplicity compared with the robot, the dedicated device is much cheaper, more reliable and more accurate.

5.6.4 *Comparison of applications*

Major applications for robots include the following:

- Material handling, loading, unloading and transferring workpieces.
- Spot-welding, electric-arc welding, electric-arc cutting and riveting.
- Machining operations such as de-burring, grinding, and polishing.
- Applying adhesives and sealants, and for paint spraying.

- Automated assembly.
- Inspection and gauging.

There are also a number of factors that influence the selection of industrial robots for any given application. These can be summarised as:

(a) load carrying capacity;
(b) speed of movement;
(c) reliability;
(d) repeatability;
(e) arm configuration;
(f) degrees of freedom;
(g) control system;
(h) program memory;
(i) work envelope/space.

It is clear that the best use of an industrial robot is for complicated activities that require accurate and incremental movements with good feedback, such as spray painting. However, from the previous discussion concerning accuracy and cost, a dedicated device is obviously more suited to simpler repetitive tasks such as pick and place. Whatever the sequence requirements, the attributes of the machine selected to do the job should be weighed up in terms of cost, accuracy, speed, reliability, safety and efficiency before they are purchased and/or used.

5.7 Supervisory control and data acquisiton

As previously stated, SCADA stands for supervisory control and data acquisition. As the name indicates, it is not a full control system, but focuses on the supervisory level. As such, it is purely a software package that is hierarchically superior to the hardware with which it is interfaced, generally via programmable logic controllers or other commercial hardware modules.

Supervisory control and data acquisition has been around for a long time, providing operators with information relating to the status and condition of the process to which it is applied. This information, while real time, tends to be the status of various valves, motors and levels. It allows a single view into a process that would otherwise be hidden or have to be summed by scanning many instruments and recorders.

Supervisory control and data acquisition systems can also produce programmed signals to control a variety of production devices such as valves, heaters and mixers. Alarms can be triggered if 'out of process' conditions are detected. And the data acquisition feature lets the operators collect and save real-time data on important process parameters and use that data in trend analyses, quality audits and management reports.

Supervisory control and data acquisition systems typically comprise a microcomputer linked to interface units, which allow the various process control devices to communicate with the controller.

Developments in SCADA capability and associated software products and technologies, have made it possible to utilise this data to provide both broader dissemination, and improvements in the quality and usefulness of the displayed and derived data.

Many SCADA systems have evolved with the addition of peripheral software packages to allow greater connectivity to other plant and enterprise systems.

An example showing the possible benefits of SCADA would be, for example, at an injection moulding or thermoforming site. A selected part of the SCADA monitoring process may be the contractual responsibility of a remote supplier, such as maintaining the hopper levels. In this case, the SCADA screen representing the hopper can be transmitted over an intranet (internal network) or the Internet (external, global network) to a display at the supplier's logistic site, allowing refilling to be predicted and optimised.

The peripheral software would regenerate the SCADA information into HTML (Web format) allowing the remote user to employ a Web browser to receive the information. It could be that several suppliers receive the information and bid online to fulfil the need. The whole production process is radically changed once an *e-manufacturing* approach is taken. Connecting suppliers, with information relating to the quality and nature of the supplied raw material, into the manufacturing process data through SCADA connectivity allows new setpoints to be quickly and automatically determined.

5.7.1 *How SCADA interacts with robots and other devices*

Supervisory control and data acquisition is an industrial measurement and control system consisting of a central host or master (usually called a *master station*, *master terminal unit* or MTU); one or more field data gathering and control units or 'remotes' (usually called *remote stations*, *remote terminal units* or RTUs). The SCADA system is used to monitor and control plant or equipment.

The control may be automatic or initiated by operator commands. The data acquisition is accomplished first by the RTUs scanning the field inputs connected to them – these remote units may be programmable logic controllers (PLCs). The central host will scan the RTUs and the acquired data is processed to detect alarm conditions. If an alarm condition is present, it will be displayed on a special alarm list. The data acquired can be of three main types:

1. Analogue data (i.e. real numbers) will be 'trended' (i.e. expressed graphically).
2. Digital data (on/off) may have alarms attached to one state or the other.
3. Pulse data (e.g. counting the revolutions of a meter) is normally accumulated (counted).

The primary interface to the operator is a graphical display that shows a representation of the plant or equipment in graphical form. Live data is shown as graphical shapes over a static background. As the data changes in the field, the foreground is updated. For example the gripper of the robot may be shown open or closed. Analogue data can be shown either as a number or it can be shown graphically. The system may have many such displays and the operator can select from the relevant ones at any time.

There are a number of elements that go to making a complete assembly cell SCADA system.

- Robot.
- The computer controller (supplied with the robot).
- FMT 2000, PLC.
- Panorama software.

The robot is the device used to perform specific tasks as programmed. The computer controller is the master component in the electrical system. All signals to and from the robot pass through the controller and are used by it to perform real-time calculations, and to control arm movement and position as well as executing program lines. Software is stored in the computer memory, located in the controller. The software interprets the operating instructions for the robot arm, and the controller transmits these instructions from the computer memory to the arm. From incremental encoders and potentiometers in the robot arm, the controller/computer receives data about arm position. This provides a closed-loop control of arm motions. The FMT 2000 is a PLC designed to process high-speed events on selected input signals. This part of the system is used to take input signals from the controller/computer and then execute an operation in the assembly cell. These operations are not necessarily complex and in the case of the cell, the FMT 2000 gives instructions to open and close pallet-clamps and start the conveyor belt. The FMT 2000 is also responsible for receiving signals back from the cell and then informing the computer/controller, which can then act on the information provided, i.e. execute the next line of program code. The final piece of the jigsaw is the SCADA software called Panorama. This basically brings all the various elements together and provides control from a single PC terminal. This is connected to the FMT 2000 and the robot controller via a ModBus link; ModBus is a communications protocol for the exchange of data across a serial link (RS232 type). It is most commonly used by PC applications, acting as a master controller to extract data from the PLC. The Panorama software can be used for the control of the system, i.e. send and receive signals from the FMT 2000 and execute program code, but it is mainly used for other functions such as:

- system monitoring and status using graphics;
- reporting and statistics;
- trends;
- control;
- historical data.

Figure 5.10 (a) and (b) show the interface using icons and Fig. 5.10 (c) shows the trends of operation of the robot cell.

5.7.2 *System monitoring and status*

The system monitoring software shows the operation of the work cell in graphical form. This enables the user to see immediately where the pallet is in the cell and what the robot is doing. With this software it is possible to monitor the status of the pallets. For example, whether they are empty, loaded, being loaded, being unloaded or being moved around the track and also whether any clamps are open or closed. To sum up, it shows all current actions graphically. There is no need to watch the robot in operation because

Fig. 5.10 *SCADA: (a) cell operation icons; (b) robot control icons*

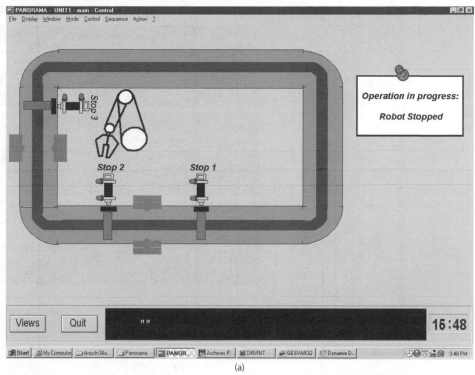

(a)

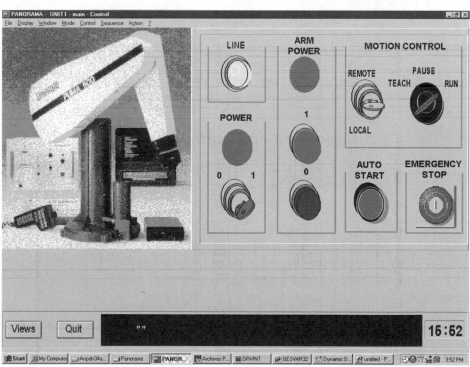

(b)

Fig. 5.10 *(c) trend charts*

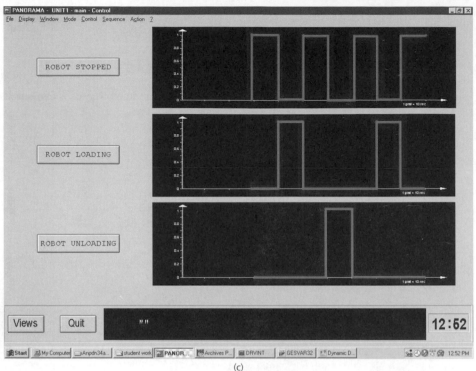

(c)

all the information is given on screen. The graphics are created specifically for the work cell so, if the cell is modified, it is possible to change the graphics to mirror the real life situation. The robot control panel is shown on the robot terminal in Fig. 5.10. This illustration shows the status of the system, i.e. whether the robot power is switched on and the mode in which it is running (teach or run), etc.

5.7.3 *Reporting and statistics*

The reporting and statistics screens presented by the software monitor the performance of the equipment and produce useful reports. The *Panorama* software allows the following information to be collected:

- number of pieces/pallets processed;
- total time taken to complete;
- total working time;
- utilisation rate of the robot (%).

This type of information is used to produce costs and estimate the output performance of the cell. It is also possible to determine the total downtime of the robot, for time and motion studies. Using older methods would have involved manual calculations.

5.7.4 *Trends*

The trend screen incorporated in the software shows graphs relating to:

- time robot stopped;
- time robot loading;
- time robot unloading.

The graphs are shown in a real time graphical mode, so the user can see the peaks and troughs of each individual action. They can be viewed over a single action or the whole process. The use of this information is very similar to that discussed in the previous section.

5.7.5 *Control*

Control, as the title suggests, is a screen that allows control of all the signals sent to the PLC from the FMT 2000 PC.

These include:

- stops: 1,2 and 3 open/close functions;
- clamps: 1 and 2 open/close functions;
- processes: i.e. start cycle, continuous cycle and stop cycle;
- robot: execute the load and unload commands.

This area gives complete manual control of the system to the user. There is also a run control in this area of the software, which allows Val 2 programs to be loaded from a disk and run. In effect, it takes away the need for the original control unit that runs of the process, however, it is still needed for the start up of the equipment.

5.7.6 *Historical data*

The software allows historical data to be stored on the computer and called back at any time. This allows users to try various operating configurations and compare the statistical and trend data. From this, best working practices can be achieved. Having this data allows users the flexibility to make minor adjustments and then see the benefits or consequences of the change; again a very useful tool.

Although the SCADA system in this section is controlling only one robot cell, in industry SCADA systems are frequently used for monitoring many cells or processes. In manufacturing institutions such as car manufacturing plants one SCADA system will monitor a number of assembly cells. The user can flick from one to the other with ease. Any problems that arise will be reported immediately and action can be taken. Information from one station can then be connected into a chain of other cells, which are all linked together to form a '*hierarchy of control*'. Theoretically this allows network monitoring of an entire plant from a single station. A simple example is now given, as shown in Fig. 5.11.

Information can be collected and collated at a monitoring station and then be passed on to a master control SCADA system. This allows a technician to monitor the performance of a whole plant at any time. This type of system does not just apply to robot assembly cells and it can be found in many industries such as:

Fig. 5.11 *The hierarchy of control*

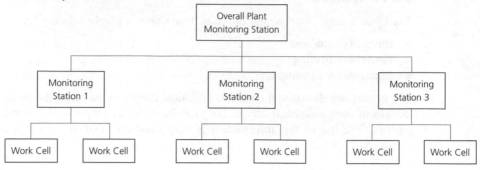

- Paper/steel/glass making (monitor each stage of the process).
- Power plant control (SCADA is used to operate fuel valves to the boilers or gas turbines, fluid control valves for steam or hydraulic systems, and circuit switching, etc.).
- Nuclear power installations (in areas where it is too dangerous for people to work).
- Chemical production (to control the flow of hazardous substances or where human error could lead to major environmental disasters).
- Bio-sensitive environments, as found in pharmaceutical manufacture, and food and drink processing (brewing and distilling), to avoid contamination by people.

Although *information control* is crucial *to industrial success*, frequently, management information is still gathered manually in most factories. Because this is slow, the data is often out of date (and therefore useless) by the time it reaches the appropriate manager. Supervisory control and data acquisition can overcome this problem. A simple package need not be expensive and the benefits it can bring in terms of providing fast and accurate information make it a cost-effective investment.

A basic SCADA system has a number of terminals on the shop floor, a central controller (usually a PC or server with adequate data storage facilities) and some form of output device such as a laser printer. Data is entered manually or with machine-readable codes. The shop-floor terminals often have a small screen to display instructions and are supplied with one of a range of *ingress protection specifications*. For example, waterproof and washable units are particularly appropriate for the food processing industries. As requirements change, the SCADA system can be expanded and upgraded to suit.

5.7.7 *Hierarchy of control*

The robot control sequence is programmed off-line using a PC based language called VAL 2. The program is then downloaded via the PC's serial port to the robot controller. The robot controller is then used as either a dumb terminal, to load the program into the PLC and display any text messages, or the terminal can be used to de-bug, edit and/or test the program.

Signals are sent from one of the robot controller's 32 outputs to the FMT 2000 PLC via its 24 volt d.c. inputs. The PLC is a sequential control system that uses 1 amp relay contacts. A series of signals is communicated to, say, a PUMA robot, which gives a feedback signal after it has completed each command.

The PC is loaded with the SCADA software and linked to the robot controller from its serial port; the serial port has 32 outputs and uses binary signals, i.e. either 0 volt or 5 volt signals, to communicate with the robot controller. The PC can run various SCADA packages, such as Panorama, and can be used to carry out various functions such as:

- *Process mimicking*, which allows the operator to interact with the process on the PC screen to provide virtual robot control. This allows the operator to drive the robot or intervene in the running of the robot. If the PC was networked the operator can do this regardless of distance.
- *Generate reports* – the reports could be anything from batch data to cycle times, downtime, number of rejects, machine tool utilisation, etc. Reporting history and trends helps to identify any areas of inefficiency.
- *Plant maintenance* – on-line plant maintenance facilities use pictures, drawings, manuals, sounds and video footage so that if the robot or process stops the operator can quickly and accurately diagnose the cause of the problem.

The schematic diagram of the robot cell system is shown in Fig. 5.12.

Fig. 5.12 *The schematic for the system of the PUMA robot cell shown in Fig. 5.8*

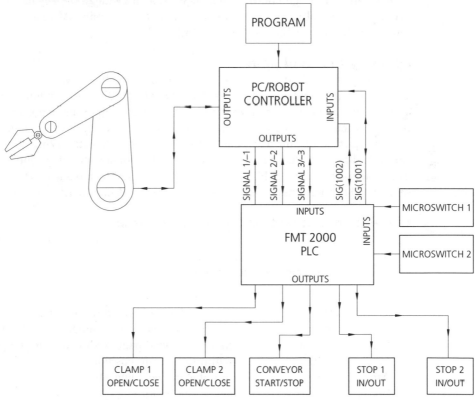

5.8 Simulation of robotic systems

There are many benefits to be gained from using graphical simulation, for example:

- Correct layout of elements in relation to one another.
- Evaluation of different models of robots.
- An opportunity to compare different operating policies.
- The prediction of performance.
- An opportunity for the education of management and operatives.
- Determination of control strategy.
- The ability to explore the above features by simulation and modelling without the need for building, disrupting the operation of or destroying the real system.
- The ability to automatically generate and test code for a variety of real robots without having to learn a specific robot language, i.e. a generic simulation language is used.

Loading the model

First the model is loaded from a library of robot models. Most robot models of any configuration and make are stored as 3D graphics with all the correct inverse kinematics joints in place. This means that the model will move in exactly the same way as the real thing. If a point in space is defined then the software will calculate how each joint will move in order for the robot to achieve that desired position. For example rules such as 'arm up' or 'arm down' can be defined along with all the angular limits that are defined by the real robot limit switches and software limits. A figure of the welding robot used in this example is shown in Fig. 5.13(a).

Defining the weld path

The weld path can be defined as an imported object or drawn within WORKSPACE as a polyline. The orientation of points along the weld path can be automatically defined as *geometry points* that all have their axes defined in the same way, for example an XY plane may be a piece of metal laid on a floor. The Z-axis is aligned with the welding torch so that the axis points vertically out of the metal. When the weld path is automatically generated, the welding torch will automatically align with these points. These points are shown in Fig. 5.13(b). They are stored within a geometry file, had they been 'taught' points they would have had to be saved in a 'teach file'.

Generating and testing the robot program

The robot program is automatically generated from an action menu and uses a generic robot language such as KAREL. Other machine specific languages can also be used. With KAREL, the commands to weld the circle are as follows:

Begin	(start the program)
Move #HOME	(move to taught home position, safe location)
Arcweld ON, sparks 100	(turn the arc weld gun on, simulated sparks 100 per min)
Follow Weld Path 1	(follow the weld path generated from the polyline)
Move #HOME	(return to the safe position)
Arcweld OFF	(turn the arc weld gun off)
End	(end the program)

Fig. 5.13 *Use of robot for welding – simulation example: (a) simulation; (b) defining the weld path*

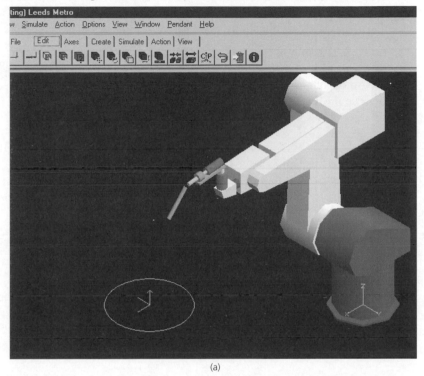

(a)

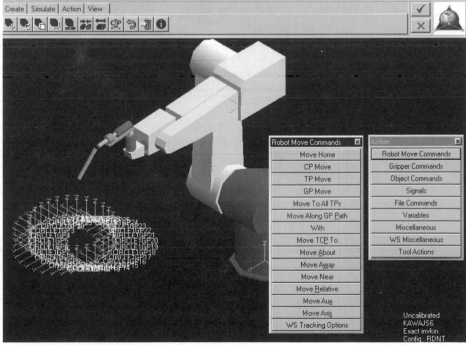

(b)

Fig. 5.13 *(c) generating and testing the program; (d) simulation of the welding operation*

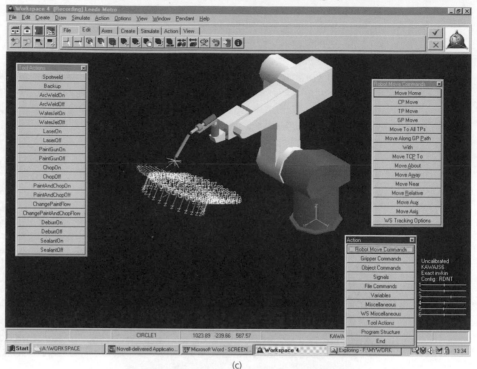

(c)

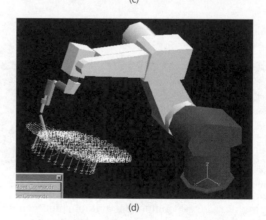

(d)

This program can then be tested to ensure that the robot can reach all weld positions and will not collide with any other object or workpiece. A test run is shown in Fig. 5.13(c). When the operator is satisfied that all is well, the program can be downloaded to an actual robot that is equipped for electric arc welding. The machine will then carry out the operation correctly as shown in Fig. 5.13(d), provided that the correct machine language from the library has been used.

5.9 Teleservice engineering in manufacturing

With the rapid development of overseas activities brought about by the globalisation of companies, it is important that these companies are able to monitor the state and progress of their plant and equipment. Service and maintenance are becoming important practices in companies wishing to maintain their manufacturing productivity and customer satisfaction overseas, in countries remote from the one in which the equipment is manufactured. The recent rush to integrate highly sophisticated manufacturing equipment utilising and supported by local design and manufacturing engineers has further increased the need that the relatively new and untested technologies collectively known as *teleservice* are able to overcome breakdown and failure problems.

With teleservice engineering the performance of a machine can be monitored and assessed from anywhere in the world. Also, information on productivity, diagnostics and service evaluation of a manufacturing system can be shared among different locations and partners.

Many US machine tool manufacturers have developed the technology for remote monitoring and diagnosis of their products. Some offer a remote troubleshooting service to their customers. The users of machine tools, for example the big US car producers, are already making use of *teleservice systems*, mainly to support manufacturing-engineering and process-development service and maintenance. The main user of teleservice for remote troubleshooting has been the office equipment industry. Many now sell their photocopying equipment with options for teleservice capabilities. Teleservice options are usually requested by customers who operate high-volume photocopying machines where reliability and throughput are important factors. Such customers are often in remote areas and their staff have only limited technical knowledge. There are a number of teleservice techniques: watchdog, intelligent tools and multimedia tools.

5.9.1 *Watchdog*

Watchdog is a technique that monitors the operational performance of components, machines and processes and can be divided into four states, namely:

1. Normal operation.
2. Degraded performance.
3. Maintenance required.
4. Total failure.

When aging occurs, the component and machine generally progress through a series of degradation states before failure occurs. If a degradation condition can be measured and detected, then a proactive and corrective maintenance activity can be peformed before a worse degradation condition or total failure occurs. A diagram of how a watchdog system operates is shown in Fig. 5.14.

5.9.2 *Intelligent tools*

This is where a watchdog system can be produced as a 'black box' monitoring unit attached directly to the machine. Every operating signature for each component is recorded and

Fig. 5.14 *'Watchdog' system for teleservicing*

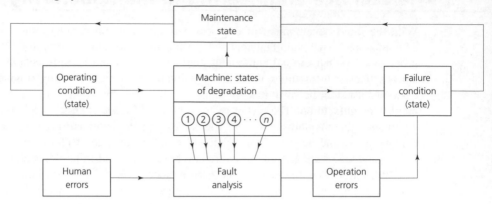

compared continuously with the required, normal-state condition. Any sign of degradation or failure can then be automatically reported. Also an operator can examine the log of the last few minutes before failure, to analyse why and what caused the problem.

5.9.3 *Multimedia tools*

Research has shown how interactive and collaborative multimedia diagnostic tools can enable technical personnel to perform the diagnosis of a problem at a remote distance. In the future *smart gloves* and *helmets* will enable maintenance engineers to perform machine maintenance and performance adjustments remotely in a virtual environment.

5.10 Group technology and flexible manufacturing systems

In Chapter 3 we examined the application of computer numerical control (CNC) to a number of manufacturing processes. So far in Chapter 5, the control and use of industrial robots have been considered. In *flexible manufacturing systems* (FMS) these two technologies are brought together to form flexible manufacturing cells. Since flexible manufacturing is a natural extension of group technology (GT), this is where we need to start.

5.10.1 *The need for group technology*

In recent years, manufacturing has had to come to terms with an ever-changing scenario in which the requirement for purpose specific, or customised, products has replaced the production of large numbers of identical products. In fact, about 90 per cent of the manufacturing base of the UK is involved with batch production rather than with continuous production.

Economically-successful batch manufacturing, incorporating modern technology, presents problems to companies whose plant layouts were probably designed many years before. Although these layouts made sense at the time of their inception, they may no longer satisfy the requirements of more modern computerised and robotised manufacturing techniques. Such techniques have been forced upon companies by fierce foreign competition to reduce unit production costs and the need to reduce the time taken to get a new product or component into production (lead time) and thus satisfy customer requirements. Furthermore, there is a constant demand for better and better quality with no corresponding increase in cost.

In response to the need for greater flexibility and a more rapid response to the demand for frequent changes in design and manufacturing methods, the *group technology* approach was first adopted in the USA and Europe during the early 1970s. It was adopted shortly afterwards in the UK. Group technology sets out to achieve the benefits of mass production in a batch production environment. It identifies and groups together similar components in order that the manufacturing processes involved can take advantage of these similarities by arranging (or grouping) the processes of production according to 'families' of components. These grouped production facilities are referred to as *cells*.

This allows different 'families' of components and/or assemblies to be manufactured using the cells almost like a number of mini-factories under one roof. These work (or manufacturing) cells have separate planning, supervision, control and even production and profitability targets. With small batch production in traditional workshops based on process layouts, the non-productive costs can be high for the following reasons:

- difficulties in scheduling;
- the transfer of work between stations that are widely separated leading to excessive handling;
- setting up production processes.

These non-productive costs as a proportion of the total cost increase as the batch size decreases. However, small batches are more easily handled by a cell because of its more closely knit and flexible organisation and the fact that, by specialising in a limited range of similar components, set-up times are reduced.

5.10.2 *Design*

New designs should make use of standard components wherever possible because this reduces costs, lead time, and ensures uniformity of quality. Where a new, non-standard component has to be used, the designer will often work from scratch rather than spend time searching a retrieval system for a similar, prior design. However, in many cases, the new drawing may turn out to be merely a variant on some previous design. There may be only a change in a single variable such as a different diameter or a change of material specification. To change the design of an existing part would be more cost-effective than starting from scratch every time.

Such an approach would require a retrieval system to be established in which all the components produced in a company would be classified and coded according to various features, for example geometric form or any manufacturing similarities. Obviously such a system would lend itself to *computerisation*. Thus the concept of a parts 'family' can be applied to the design process in addition to the manufacturing process. If fact it should, ideally, start in the design process.

5.10.3 *Parts families*

As previously stated, group technology seeks similarities not differences. Therefore, all similar parts can be collected into families. A 'family' or 'parts family' is a collection of components that have similarities in geometric form or are manufactured by similar manufacturing methods. Once the family has been identified, then a composite component can be drawn up that contains all the features of the parts family.

The available process machines can then be sorted out into the best mix to produce the composite component. The group of machines selected is then physically moved together to create a *work cell*. The machines in the cell are then set up to produce the composite component, the actual production components are manufactured by omitting those operations not needed on any particular component.

5.10.4 *Management*

With conventional process layouts for batch production, production control increasingly demands more up-to-date information. This has led to some very complex control systems that attempt to provide management data on a day-to-day basis and even, in some extreme cases, on an hourly basis.

Managing a group technology system can benefit the information gathering process since each group runs as a separate entity and it is only necessary to plan the work into and out of the cell. There is no necessity to plan work through each machine since each cell is self-monitoring. Group self-control is the biggest and most difficult change for conventional management to accept, since responsibility and, therefore, authority is placed firmly at the point of production. External interference by the detailed planning of each operation would effectively destroy the concept of group technology.

5.10.5 *Flexible manufacturing systems*

As previously stated, the natural extension of group technology is to move to manufacturing cells in which all the functions are externally *controlled by computers*. That is, a 'people-less' manufacturing cell. In group technology the human beings are part of the process of manufacture but in flexible manufacturing systems the human effort is confined to component and tooling design, setting up the machines and equipment in the cell, and providing high level maintenance expertise. It is even possible to install machines that can change their own tools and set themselves.

Again there is a need for parts-families but, in the case of flexible manufacturing systems, the cell can handle much larger families of parts and deal with them in random order. Other differences are:

- the need for transport (mechanical handling) equipment that can move components from machine to machine in the correct sequence and maintain maximum loading for each machine;
- a computer that can control all the operations and materials handling at the same time.

The outstanding difference between a GT cell and an FMS cell is that the latter involves a very much higher level of capital investment. Before investing in an FMS cell,

Fig. 5.15 *Position of FMS in the production hierarchy*

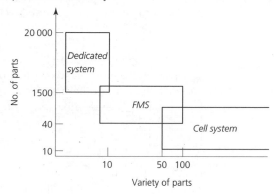

comprehensive feasibility studies have to be carried out and these must include a five-year market forecast, an estimate of the return on the capital invested and the level of flexibility required in the system.

Basically, FMS is a group of computer numerically controlled workstations interconnected by means of a materials handling system. The individual computers of the workstations and the handling system are themselves under the control of a *master computer*. A hierarchy of control exists in which the master computer deals with the management of the system such as the scheduling of the parts. A *supervisor computer* deals with monitoring the system and acts as the interface between the system and the operator. For example, error conditions are reported and a graphical interface or SCADA system shows the current state of each part, i.e. where it is within the system, if a machine is busy or if it is waiting for work. In terms of production volume, FMS lies between the high-volume production of dedicated machines used for mass production and individual CNC machines, which are best suited for batch production in a group technology cell where there is a large variety of products. This is shown in Fig. 5.15.

5.10.6 *Concepts*

Where the parts manufactured on flexible manufacturing systems are prismatic in shape, they can be loaded into pallets carrying the necessary fixtures and transported on conveyors or by automated vehicles whose path is determined by *guideways* in the workshop floor. Such systems are referred to as 'primary conveying systems'. The pallets are coded with a bar code carrying the machining instructions. The bar code is 'read' by sensing devices, which route the part to the particular machining destination or to a buffer store conveyor until the required machine is available. Thus, while the part is within the system, it is either in transit or being machined. The system has secondary conveyors whose function it is to receive the parts from the primary system and transport them to the individual machines or buffer stores.

The need for buffer stores is due to the different machining times for each operation. Thus buffer stores allow short operations to be accommodated with those requiring longer machining times. The part remains within the system until all the operations on it are complete. Remembering that this is a flexible system, different parts requiring

Fig. 5.16 *Typical FMS cell*

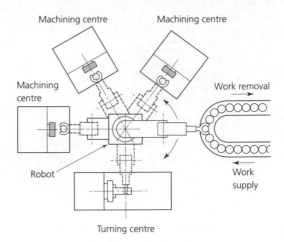

different operations can be accommodated within the system at the same time provided that the tooling is available. Since the development of *industrial robots*, the range of components that can be handled has grown beyond those with simple prismatic geometry and the overall flexibility has been enhanced. An example of a typical cell for a flexible manufacturing system is shown in Fig. 5.16.

5.10.7 *Techniques available for determining FMS layout and control*

Current trends in the layout design and control of FMS have turned towards simulation and modelling. An indication of the importance of simulation and modeling is illustrated by companies such as Citroën and Yamasaki, with the latter company spending 100 000 man-hours of planning prior to installation. Both companies reported that they derived considerable benefits from using computer simulation. Let us examine the advantages that can be derived from any technique that can model such a system. For example, modelling could achieve the following:

- Correct layout of elements in relation to one another.
- Evaluation of different designs.
- An opportunity to compare different operating policies.
- The prediction of performance.
- An opportunity for the education of management and operatives.
- Determination of control strategy.
- The ability to explore the above features by simulation and modelling without the need for building, disrupting the operation of or destroying a real system.

Most systems are subject to random influences, such as the vagaries of human beings; however, since a flexible manufacturing system is computer controlled, many of the random disturbances can be eliminated. This means that there can be greater confidence in the results of a flexible manufacturing system over conventional manufacturing methods.

Fig. 5.17 *GRASP FMS cell*

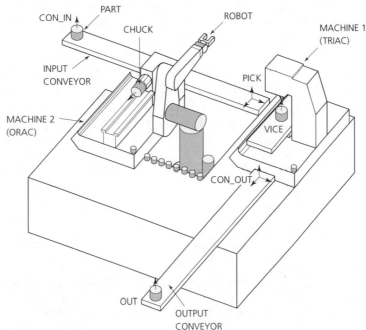

5.10.8 *Graphical simulation*

This is where the physical design and interaction between elements within an FMS can be visualised, and it has the advantage of giving the system designer information so as to identify excesses and deficiencies within the system and, to some extent, predict how the system will perform. Examples of 3D design packages are GRASP, Workspace and RobCAD. They have all been developed to model robots, conveyors, automatic-guided vehicle systems (AGVS), etc. Such features aid the development of workplace layout by manipulation of these entities. Other features include *event processors* that can link events together such as robots picking and placing objects on a conveyor. Also CNC machines can be automatically started and stopped. These events can then be saved in the computer memory along with their associated parameters, e.g. starting time, duration of actions, delays, accelerations. In this manner complex interactions between components within an FMS may be studied and evaluated. An example of a typical cell containing a robot, conveyors, a CNC lathe and a CNC milling machine is shown in Fig. 5.17.

The following example is based upon the flexible manufacturing cell shown in Fig. 5.17 The GRASP sequence of a part passing through the system is as follows:

- PART enters on the INPUT conveyor to a sensor at PICK
- from PICK to VICE
- PART processed on the TRIAC milling machine
- from VICE to CHUCK
- PART processed on the ORAC lathe
- PART exits on the OUTPUT conveyor at CON_OUT

Fig. 5.18 *Full GRASP program*

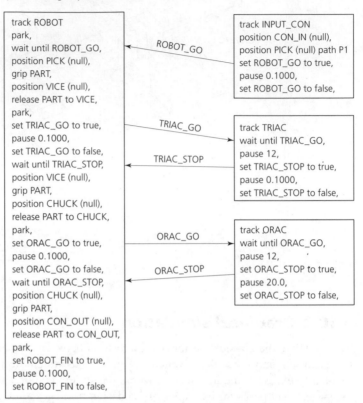

global variables
signal ROBOT_GO, ROBOT_FIN, TRIAC_GO, TRIAC_STOP, ORAC_GO, ORAC_STOP;
path P1 straight speed 500.0000;

```
track ROBOT
park,
wait until ROBOT_GO,                          track INPUT_CON
position PICK (null),          ROBOT_GO       position CON_IN (null),
grip PART,                                     position PICK (null) path P1
position VICE (null),                          set ROBOT_GO to true,
release PART to VICE,                          pause 0.1000,
park,                                          set ROBOT_GO to false,
set TRIAC_GO to true,          TRIAC_GO
pause 0.1000,                                  track TRIAC
set TRIAC_GO to false,                         wait until TRIAC_GO,
wait until TRIAC_STOP,         TRIAC_STOP      pause 12,
position VICE (null),                          set TRIAC_STOP to true,
grip PART,                                      pause 0.1000,
position CHUCK (null),                         set TRIAC_STOP to false,
release PART to CHUCK,
park,                                          track ORAC
set ORAC_GO to true,           ORAC_GO         wait until ORAC_GO,
pause 0.1000,                                  pause 12,
set ORAC_GO to false,          ORAC_STOP       set ORAC_STOP to true,
wait until ORAC_STOP,                          pause 20.0,
position CHUCK (null),                         set ORAC_STOP to false,
grip PART,
position CON_OUT (null),
release PART to CON_OUT,
park,
set ROBOT_FIN to true,
pause 0.1000,
set ROBOT_FIN to false,
```

A graphical robot and simulation package (GRASP) can be used to write a programme for this sequence. The GRASP commands are similar to the VAL 2 programming language, i.e.

POSITION vice shift Z100	means position the TCP 100 mm above vice
GRIP part	means grip the workpiece part
RELEASE part to chuck	means release the part to the location CHUCK
SET ROB_GO	means set the signal ROBOT_GO to *true*
WAIT until TRIAC_STOP	means wait until the signal TRIAC_STOP is *true*

The full GRASP program for the previous sequence is shown in Fig. 5.18.

5.10.9 *Discrete event simulation*

In a discrete model the states of each *entity* within an FMS are modelled individually, for example a machine working or waiting, or a workpiece waiting or being worked on. There are combined activities such as machine–process–workpiece activity. In other states entities can be in a queue waiting for conditions to change, for example a workpiece waiting for a machine to become available. The selection of an entity from a queue depends on its characteristics, such as the type of machine required. Once the entity has been selected its state is changed, e.g. the machine is working. The activities each entity undergoes are considered to begin and end instantaneously and are known as 'events'. Generally these types of simulation are controlled by a timing mechanism known as the 'three phase' method, for example:

- advance the clock to the earliest event;
- terminate any activities that are due to finish at that moment;
- initiate any activities permitted by the conditions built into the model;
- repeat the process.

From this type of model a variety of performance measures can be considered such as machine utilisation, work in progress (WIP), throughput times, buffer queue performance data, etc. An example of a discrete event package is PROMODEL and SIMAN and HOCUS are examples of modelling languages.

5.10.10 *Relationships between parts and machine tools*

The need for a parts family is at the root of FMS. Thus there has to be a close liaison between the manufacturing engineers and the designers in order that the machining data can be set and agreed. The location points for the fixtures are of equal importance, together with the state of the raw material entering the system. All these facts have to be agreed and established at the planning stage.

The part size will influence the choice of machine within the system (large part: large machine) and the part geometry determines the type of machine tool, as it does in general manufacturing. Where the variety of parts to be manufactured is large, then standard CNC machines should be considered. However, the larger the volume and the smaller the range of components to be manufactured, the greater the tendency to move towards dedicated machines and equipment. This will reduce the flexibility but increase the rate of production. Care has to be taken when considering the part/machine-tool relationship to keep future production requirements in mind. If the market is fluid, then flexibility is the key factor but if the future demand is more stable, then less flexibility and greater production rates will be required.

The variety and complexity of the workholding fixtures in FMS make great demands on the ingenuity of production engineers. As in all machining operations, the accuracy of the final product is only as good as the accuracy of presentation of the work to the cutting tools. The concept of the parts family leads to the identification of a common datum and common location features. The object, as in all jig and fixture design, is to reduce loading time. This is particularly important in FMS, where gains in production times can be easily offset by loading costs. Inspection time and costs must also be reduced if the full benefits of FMS are to be achieved. Automatic inspection stations or probes set within the machine tools themselves must be employed. Actual dimensions are compared with standard dimensions by the computer and adjustments are automatically made to the machine tools. Waste material (swarf) will be produced in large quantities, so automatic swarf handling equipment has to be considered. All too often, high-investment systems can be brought to a halt as tools become broken and movements clogged, simply because adequate swarf removal facilities have been overlooked.

5.10.11 *Materials handling*

Materials handling is the element in any FMS cell that builds flexibility into the system. Its functions are to move the work between machines, buffer stores, inspection stations, etc., and also to present the work, correctly orientated for cutting, to the machines. The handling system must have the facility to independently move workpieces between machines. That is, such workpieces must be able to flow from one station to another as the loading and routing demands. There must also be buffer storage facilities for work waiting to be machined.

At the machine tool, the secondary handling systems must permit workpiece orientation and location for clamping within the workholding fixture ready for machining. For geometrically-simple, prismatic parts families, palletisation on conveyor systems or automatic guided vehicles (carts) is the most common primary transport system. Pusherbars, guideways or robots transfer the parts between primary and secondary systems and load the work stations.

The development of industrial robotics has increased the flexibility and versatility of flexible manufacturing systems by enabling rotary parts to be included. The robots can transfer parts between systems and load them into machine fixtures under the control of the master computer. A single robot can be installed in the centre of the system and can load each machine with parts at random as the routing requires, see Figs 5.16 and 5.17 shown previously. It will, of course, also unload the parts when the process is complete. The machine tools must be laid out so that they are within the operating radius (reach) of the robot's arm. The *end effector* that is mounted on the 'wrist' of the robot must be able to hold all the components within the parts family of the cell.

5.10.12 *People/systems relationships*

While the system itself is automatic, people still have a role to play. This role is one of system management rather than machine operating. The supply of raw material to the FMS cell and the removal of the finished components are still manual operations. In the long term completely automated factories could become feasible but not necessarily desirable.

The major change is in the content of the work available. The need for trade skills is reduced, but more emphasis is placed upon office-based skills, and upon design and production-engineering skills. Sociologically the implications are profound with a lessening in demand for skilled and semi-skilled operators, a shortening in the working week and the virtual elimination of the need to work 'unsociable' hours. Robots and computers do not have to see, so round the clock 'lights out' operation of the factory can lead to substantial savings in operating costs. The manning needs for such a system are for technicians and engineers of the highest skills in the fields of product and tool design, using CAD/CAM techniques, tool setting and changing, and for multiskilled maintenance engineers with expertise in electrical, electronic, mechanical, hydraulic and pneumatic systems. The computer system makes its own demand on programmers and system analysts.

5.10.13 *Benefits*

Having examined the technology and management of FMS cells, let us now consider the benefits of such a system. The manufacturing sector is essential to the UK economy. It absorbs 20.2% of the employed population, provides 80% of all UK exports and accounts for 20% of the gross domestic product (GDP). Despite its importance to the economy of the UK, the majority of jobs in this sector of industry consist of manufacturing a great variety of parts in low volumes; a figure as high as 80% with batch sizes of 10 to 50 units is typical. Also the market life of certain products, for example telephones, has been reduced from 10 years to 18 months. This makes the need to respond quickly to market needs more important than ever and has confirmed that FMS has a place in allowing the implementation of changes from old to new designs, with low investment costs. This view is supported by *Masuyama of Toyota* who states, 'it seems essential to perceive accurately the condition of the market and to supply what is demanded by the market, with a short lead time and at a low cost'.

Companies that have installed FMS talk of reducing lead times by 50% to 60% and the reduction of scrap levels from 25% to 5% of output. Companies who invest in FMS are looking for a pay-back period of two to three years. Also FMS promises productivity improvements through increased machine utilisation and reduced levels of work in progress, along with a reduction in production cycle time.

An FMS system provides the flexibility needed to make a large variety of components on a continuous basis. The cost of production when using FMS can be 50% lower than the cost of conventional methods and production can be increased by a multiple of 2 to 3.5. This shows clearly that FMS must continue to be developed and replace conventional methods of manufacture if the manufacturing industry in the UK is to survive in competitive markets. It should now be apparent that the adoption of FMS strategies has many advantages and we can summarise the commercial benefits as follows:

- A faster response to market changes in design or the creation of new designs.
- Improved product quality.
- Direct and indirect labour costs are reduced.
- Production planning is improved.
- Machine utilisation levels are improved.
- The 'work in progress' inventory is improved since the time spent in the systems is reduced.

It is not possible to buy a ready-made flexible manufacturing system, each installation has to be tailored to suit specific company needs. Therefore once a company has decided to follow the FMS route, it has to look at every aspect of its production practices, both good and bad. It is in this analysis and during pre-planning that the greatest benefit to the company occurs. Even if the decision is made not to invest in FMS, the feasibility study will have revealed flaws within the present methods of production that can be corrected and the company will benefit. Also, the level of management commitment, so essential in the implementation of FMS, will have been explored.

5.10.14 *Problems associated with FMS*

The many benefits accrued by installing FMS are not easily achieved, and there are many problems to overcome before the full potential of the system is realised. The main problems associated with the installation of FMS are:

Cost

The capital investment of an FMS is very high, ranging from £100 000 to many millions of pounds. Therefore poor selection criteria and implementation methods can produce costly mistakes.

Layout

The physical relationship between entities within an FMS is critical. For example matching the working envelope of a robot so that all the target positions can be reached within the FMS is very difficult. Also, determining the configuration of the end-effector (gripper) to a work-holding device can be very complex. Other physical problems encountered include matching the velocities of components on conveyors, automatic guided vehicles (AGVs) and robot arms.

Efficiency

The development of efficient control strategies for the maximum utilisation of an FMS system is most important but, at the same time, extremely difficult. The large capital investment in an FMS cell necessitates the maximum utilisation of the system if an acceptable return on investment is to be made. A survey of 13 Swedish companies indicated that FMS has increased machine utilisation rates by an average of 60% and reduced work in progress and 'lead time' by a similar amount. Typical system capital costs ranged from £0.8M to £3.0M in 1996. The payback periods ranged from 2 to 5 years.

Flexibility

The dichotomy of achieving part variety and flexibility while, at the same time, maintaining efficiency should be apparent. It is clear that flow-line production must be at its most efficient if only one component is being manufactured, since it enables all processes to be fine tuned to achieve maximum efficiency. At the same time, by its very nature, flow-line production has little flexibility. On the other hand, FMS has much greater flexibility, and it is because of the variety of parts manufactured in different batch sizes and at different intervals that the control strategy of FMS is both complex and difficult to plan. For example, how is a critical component to be made ahead of schedule? This will affect the control strategy and efficiency of the system and will require the scheduled master computer to be reprogrammed.

5.11 Measurement and quality control

No manufacturing system, however cleverly automated by the use of information and communications technology (ICT), can function effectively without equally effective systems of measurement and quality control. There are many systems of measurement both manual and electronic in use in manufacturing industry; however, in this section only three-dimensional coordinate measuring machines will be considered since they are used increasingly in conjunction with FMS cells.

5.11.1 *Three-dimensional coordinate measuring machines*

The three-dimensional coordinate measuring machine (CMM) is now used extensively in the manufacturing industries as a powerful and versatile measuring machine. It is an indispensable tool used in quality control departments that pursue increased efficiency and accuracy in dimensional, geometric and contour measurements. The CNC controlled coordinate measuring machines are playing an increasingly important role in flexible manufacturing systems.

The coordinate measuring machine may be defined as a 'machine that employs three movable components that travel along mutually perpendicular guideways to measure a workpiece by determining the X-, Y-, and Z-coordinates of points on the workpiece with a contact or a non-contact probe. The position of the probe at the moment of contact is determined by displacement measuring systems (scales), associated with the three mutually perpendicular axes'. Since measurements are represented in a three-dimensional coordinate system, a CMM can make many different types of measurements such as dimensional, positional, geometrical, deviation and contour.

Measurement and inspection work using conventional measuring equipment and gauges requires expertise and experience in order to ensure that reliable measurements are obtained with high repeatability. Not only is availability of this expertise and experience becoming increasingly limited and costly, the increased use of CNC machining has resulted in the economic manufacture of increasingly complex components. Conventional inspection techniques can no longer be used to check complex-shaped components involving curved surfaces and imaginary origin points. These problems have been eased to a large extent by the increasing availability and use of CMMs.

Three-dimensional coordinate measuring machines may be manually controlled or CNC controlled. Figure 5.19(a) shows a typical manually operated coordinate measuring machine. Such machines only became possible with the introduction of the touch trigger probes

Fig. 5.19 *Manual coordinate measurement: (a) manual coordinate measuring machine; (b) configuration of the manual system*

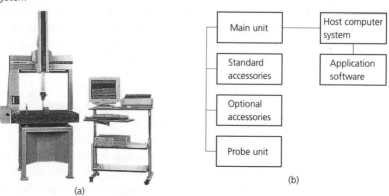

(a)

(b)

Fig. 5.20 *CNC coordinate measurement: (a) CNC coordinate measuring machine (CMM); (b) configuration for CNC system*

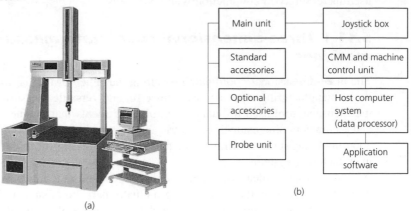

(a)

(b)

introduced in the previous section. The system configuration is shown in Fig. 5.19(b). This CMM, being a manually operated machine, is driven by the operator by means of a joystick. The resolution of this machine is 0.0005 mm. To ensure smooth and accurate movement air bearings are used on all three axes and the Z-axis is mass balanced.

Coordinate measuring machines controlled by CNC are used to maximise the efficiency of automated manufacturing processes that demand fast, efficient and accurate measurement. The digital servo controller automatically seeks the most efficient, and fastest, probing path. An example of a typical CNC coordinate measuring machine is shown in Fig. 5.20(a). This machine can be CNC operated or driven manually by means of a joystick. The system configuration is shown in Fig. 5.20(b). The resolution of this machine is 0.0005 mm. To ensure smooth and accurate movement air bearings are used on all three axes and the Z-axis is air balanced.

Let us now consider the basic principles of using a manually operated machine. The component to be measured is mounted on the table of the machine. No clamps are used, so as to avoid distortion. The movement of the probe along the X-, Y-, and Z-axes is achieved by a simple joystick. When the probe touches the workpiece surface the machine stops no matter what the position of the joystick so as to avoid damage to the probe (touch-stop function). The position of the probe is fed automatically into the computer. The system also prevents further movement in the previous direction (restricted area) and the probe must be withdrawn before it can be moved to its next destination. To establish the position of the workpiece on the machine table, the probe touches on at least two points on each of three mutually perpendicular faces as shown in Fig. 5.21(a). These positions are inputted directly into the computer. This automatically assesses the position and alignment of the computer on the worktable and establishes the included corner as a point datum as shown in Fig. 5.21(b). This system avoids the need for elaborate work-holding fixtures to position all the components in a batch at exactly the same position on the worktable as each one is measured. The probe is now moved to the next measuring point *P* (Fig. 5.21(c)). The computer will display the position of this point relative to the origin. By moving the probe to each of the pre-planned, key inspection points on the component in turn, all the dimensions can be checked.

Fig. 5.21 *Principle of use of a coordinate measuring machine: (a) touch probe on position T_1–T_5 inclusive; (b) computer identifies point of origin and alignment of component from the probe signals received in (a) at the common corner for the three faces A, B and C; (c) touching the point P will determine the distance x from the face C*

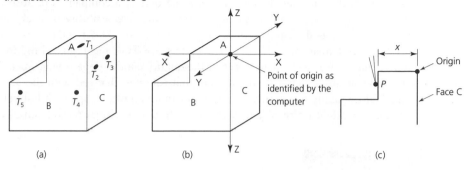

Where a large number of components of the same type are to be inspected, a CNC machine can be used. This works in exactly the same way except that the controller is programmed to carry out all the movements necessary to set the point of origin and inspect the components. The components can be loaded onto the machine table by conveyor or robot. This enables the CMM to be linked into an FMS cell for lights-out, unmanned production. Automatic probe changing systems are also available with CNC operated CMMs adding still further to their versatility. The computer can be programmed to compare the readings from the CMM and probe with the ideal dimensions for the component and thus determine any errors that may be present and their magnitude. It will also print out a certificate of conformance for the component under inspection.

5.11.2 *Quality in manufacture*

In this chapter we have considered the control of manufacture in some detail. We have also considered the application of computerised techniques to measurement and inspection. The final link in the chain to achieve world class manufacturing is quality control to international standards.

The word 'quality' is bandied about quite loosely these days, especially in publicity material. As applied to manufacturing it is defined in BS 4778 (1987) as '*the totality of all features and characteristics of a product or service that bear upon its ability to satisfy stated or implied needs*'. Put more simply it can be defined as 'fitness for purpose'. The survival of manufacturing companies in the international markets depends upon the achievement of acceptable levels of quality, at minimum cost to the customer. There is no such thing as absolute quality. A customer's idea of quality will change with time and with competition in choice. Therefore a company's attitude to quality must be constantly reviewed in light of these changes in customer perceptions so that, at all times, that company's products represent 'fitness for purpose', 'value for money' and 'state-of-the-art technology'.

The price of any commodity is negotiable; quality is definitely not negotiable. This is widely appreciated in industry because most manufacturers are, themselves, customers in one way or another. Any apparent savings made by the purchase of inferior materials or components are quickly dissipated in loss of customer confidence. The quality of goods or services goes hand in hand with reputation. This goodwill is a company's greatest asset. It does not just happen, it has to be part of a company's philosophy. That is, it has to be managed so that quality becomes the responsibility of every member of a company. This is the concept of *total quality control* (TQC) and its implementation is total quality management (TQM). The concept of TQC has replaced the former system where quality was the sole preserve of the quality control department. Such departments still have an important role to play, but that role has changed with the advent of TQC and TQM. Quality is not only an important function of management but being 'systems driven' it lends itself readily to computerisation and will be considered in Chapter 6.

ASSIGNMENTS

1. Visit the www.coltergroup.co.uk Web site and download a free trial copy of their offline programming software. Try and copy the ladder diagram and instruction set code into this system from the pick and place example in this chapter.

2. Research the different types of PLC manufacturer and determine if their controllers can network with each other, if they can be programmed offline and if they can be interfaced with a SCADA system.

3. Design a simple schematic diagram of how you imagine the distributed control of a car manufacturing plant would be.

4. Now determine the sort of information that would be displayed to an operator and the type of data that would be logged for further analysis, what might that analysis be?

5. Visit the different Web sites for robot manufacturers and determine what sort of system they can interface to. Also whether they can be programmed offline and any other special features they may present. Some examples of manufacturer include:
    ● www.staubli.com
    ● www.adept.com
    ● www.cincinnati.com
    ● www.epson.com
    ● www.abb.com

6. Visit the workspace robot simulation Web site and determine the different types of robot that can be programmed and simulated using their software.

7. Research other simulation Web site software and determine the different processes that can be simulated.

8. Discuss the principles of, and need for, component grouping when considering the economics of flexible manufacturing systems.

9. Discuss the economic and technological criteria that need to be considered when analysing the need for, and benefits of, setting up a flexible manufacturing cell.

Part C
E-management

6 Developments in organisation and management philosophies

When you have read this chapter you should be able to:

- appreciate the effects the Internet and intranet have had on industry;
- understand the benefits of e-commerce and on-line communities;
- appreciate how hypermedia and multimedia are used;
- describe the management of manufacture;
- appreciate the concept of world-class manufacture;
- understand the benefits of MRP 2 and Just in Time;
- appreciate the benefits of:
  - computer integrated manufacture;
  - computer aided process planning;
  - computer aided production management;
  - product data management systems.

6.1 Introduction

This chapter discusses the impact of Internet technology on the management of engineering and, in particular, manufacturing. It suggests how this technology may influence future computer aided engineering (CAE) operations and aid communications between both internal and external engineering clients. It also highlights how a company's intranet can be used to disseminate information across a corporate site via a local area network (LAN). In this chapter we will examine customer relations, intranet and documentation, education and training and marketing and commerce.

The *Internet* is a term used to describe an interconnection of worldwide computer networks, and as such it is not one entity, but a cooperative system of networks linking many institutions and organisations. The World Wide Web is a term used to describe the hypertext interface to the information available on the Internet. Hence any computer with a link to a host server will have the ability to link to any document held on any host server in any location around the world. In December 1998, 12.5 million European households had access to the Internet. Predictions are that the number could exceed 43 million by the year 2003. For any organisation, the Internet provides a way to communicate with external customers and suppliers.

6.2 Customer relations

Customer–supplier relations have undergone vast improvements owing, in part, to the implementation of technology. There has been a steady increase in the number of companies using on-line ordering systems. This has been both internal and external. The development of manufacturing resource planning (MRP) systems has gone a long way towards optimising internal customer–supplier relationships and, in more recent years, even small concerns have realised the benefits of communication by the use of e-mail and the Internet.

Real time communication is now possible between individual departments situated within a company. Orders can be placed and confirmed in less time than was previously possible. Using MRP with 'Just In Time' (JIT) techniques allows full optimisation of the manufacturing process and close control of workflow and internal inventory levels. Adopting the JIT philosophy ensures that supplies arrive just in time and in the correct place to be used immediately so that there is no excess stock tying up factory floor space and working capital funds. The MRP and JIT systems can be extended to include external suppliers and if they are operating compatible systems, integration between the two can be accomplished.

External customers benefit from the improved communications offered by technology, by being able to maintain close contact with their supplier. In the period during the design of a product, images and drawings of the developing design can be transferred between the various parties for evaluation and modification in a very short time. The progress of a project can be monitored by frequent update reports generated automatically by the supplier's system. Furthermore, dispatch and expected delivery dates can be advised as they occur.

Possibilities include 'real-time' teleconferencing, an area undergoing considerable development. As the speed of speech and graphic data transmissions is improved, virtual interaction will become a viable everyday tool. Advances in 'wireless Internet' are also beginning to have an impact. Amateur radio operators have been using this technology for several years, albeit in a primitive form. The advances will allow communications from remote locations to become a norm, rather than an advanced technology.

6.3 Intranets and documentation

With the sheer volume of information stored by national and global corporations – including reams of printed information such as computer documentation, procedures, specifications and reference documents – the argument for making *information on-line* available becomes increasingly valid.

Users no longer have time to find some obscure item of information from a shelf full of manuals. Companies can no longer justify the cost of printing all this information without any guarantee that users are actually going to read it. There is also the problem of keeping all printed information up to date. Many corporations are already using an intranet to deliver information to internal users. The term *intranet* refers to the fact that this internal web is run inside a private network, often without a direct connection to the Internet (the external Web). There are a number of information resources and transactions that are potential candidates for an intranet.

Every corporation has reams of business information that it must distribute to internal employees or external customers and suppliers. The following list provides examples of the types of documents corporations can distribute over an intranet:

- seminars;
- policy and procedure manuals;
- company newsletters and announcements;
- scheduling information;
- map and schematic drawings;
- computer reports;
- customer data sales and marketing literature specifications;
- price lists;
- product catalogues;
- press releases;
- quality manuals;
- ISO 9000 quality control work instructions;
- employee benefits programs;
- orientation materials;
- software user guides;
- hardware manuals;
- quick reference guides;
- on-line help style guides and other standards, training manuals and tutorials.

An intranet allows corporations to put all documentation into one source (database). This source is constantly updated to ensure that users always get up to date information. Information stored on an intranet is much easier to control than in the past. Before intranet, many copies of the same documents were kept in a company's departments. Inevitably, many of the documents became out of date but were not replaced.

6.4 Education and training

The delivery of education and training is shifting from the traditional methods of colleges, universities and private training specialists to wider and more open systems. Technology is enabling 'distance' learning to develop to a level previously unknown. Students can now use interactive materials from remote locations while remaining in close contact with their training and academic providers, through use of the Internet and global e-mail systems. While, ideally, there is no substitute for a 'one-to-one' learning relationship with a tutor, there is a definite move towards technology-based training. This ensures that students, apprentices and established staff are able to study and keep up to date in their individual fields of activity, yet have the geographical flexibility that industry demands. For example, a student engineer working on an oil-rig cannot keep returning to the mainland to attend lectures and tutorials but can keep up to date with his or her studies via a range of communication systems and technologies, as stated below.

This training can be delivered in a variety of formats, i.e. interactive CD-ROMS, Web sites, intranets. A growing number of university courses are offered 'on-line'. For example, in the field of engineering systems and design, a growing number of universities offer training packages that can be accessed by anyone with an Internet connection. Feedback on student results is limited, however, because the 'one-to-one' tutorial is only available to individuals registered with a given provider.

Nonetheless, the potential of such methods to deliver education and training is becoming recognised and those who can provide these on-line courses will become a valuable resource. The training of personnel, in a professional environment, often results in disruption to a company's daily activities due to the need for the student to attend some external establishment in order to receive the relevant training. Technology allows the training to be delivered and undertaken directly at the student's or trainee's place of work. This minimises the need to 'release' a given employee for more time than is absolutely necessary.

The delivery of training materials is likely to include video-conferencing techniques, thereby reinforcing the bond between student and teacher and providing a more personal relationship that encourages the student to ask questions more frequently. This bonding is an important part of the learning process and, unfortunately, is an element missing from the majority of current 'interactive-on-line-training' regimes.

Research in the field of training shows that, in the future, it will increasingly be undertaken on-line. For the training provider the implication is that of higher internal efficiency, reduced cost and the elimination of the need to provide a location where students can study. The student has to source the extra materials required to complete their study. This will result in only the best students attaining top grades, due their dedication and autonomy and, most importantly, a natural filtration of those studying under peer pressure to 'achieve' in some field to which they are not entirely committed.

From the foregoing, it should be evident that education via a corporate intranet would mean that employees should be able to undergo training and study for qualifications without leaving the workplace. Corporations spend thousands of pounds each year sending their employees on external training courses and to university. The greatest cost to the company is not the course fees, it is the time the student is away from the workplace when company policy is to pay employees for time spent on training as if they were at work. Education and training via a corporate intranet and the Internet would drastically reduce the cost of training an employee. For example many corporations incur costs of around £30 000 in lost work while the average day release student is off-site.

6.5 Marketing and commerce

As a resource for marketing and market research, the Internet provides a wide and diverse pool of information. Advances in intranet security have resulted in the development of secure *e-commerce* and the ability to conduct financial transactions in relative safety.

Companies can benefit from both of these areas by using the Internet for advertising and research purposes, and for selling their own particular products. The Internet also provides access to on-line databases, some of which charge per search and some of which are free.

Companies and individuals can source specific information about competitors and potential suppliers. Many large companies publish their annual reports on the Internet thereby providing essential information for investors. As a resource for market research, Internet technology has increased the available information manyfold. This is not without its problems, however, because the diversity of the available information often results in simple searches returning large amounts of unrelated information. Best results are usually obtained by using *specialist search engines* and/or *advanced Internet search criteria* that are able to filter out the irrelevant information.

E-commerce is a growing area on the Internet. Many companies (and private individuals) now provide secure sites where transactions can be undertaken. For example, quotations requested for a number of high-end graphics workstations for running a company's CAD systems were sought and provided directly on-line and to the company's own exacting specification. Traditionally, this would have entailed writing or faxing the specifications to the suppliers concerned. They would then have had to generate the quotation and return this to the buyer. Using the Internet the transaction was completed in minutes, rather than the normal wait of several days using surface mail.

This vast array of commercial information on the Internet has opened areas for business that were hitherto unavailable. The resulting data traffic that has been generated by the success of the Internet has meant advances in Web technology and access to wide bandwidth transmission carriers in the form of the 'superhighway' and ISDN. The ultimate in transmission speed is obtained by 'hard-wire' links, leased from the communications companies, to the Internet server. This is common practice in the USA and it is becoming more frequent in other technology-based locales.

There is a challenge for the suppliers of technology to keep pace with the growing needs of the Internet communities as the on-line business they undertake increases. The scope for this is limited only by the available technology. Currently, marketing is the main focus on the Internet. This includes:

- product brochures;
- on-line electronic catalogues;
- product specifications and manuals;
- product promotions and discounts;
- real-time product pricing;
- ordering information and billing;
- lists of sales contacts and multimedia demonstrations.

6.6 E-commerce

Originally, much information on Web sites has been read-only, but the developments of Web site technology, mainly JAVA, now mean that Web sites are likely to become interactive. Such developments in e-commerce will result in the entire sales and marketing process being taken on-line. If companies are to succeed in the world of e-business, they will need to re-orient their entire transactional model towards the Internet. The Internet has helped many companies make their products available globally. Its introduction has put products into the marketplace much more quickly. It could mark the end of 'cold-canvassing' for the old style sales representative. Initial contact will be made via the Internet and interactive Web sites with virtual three-dimensional products. A personal visit from a technical representative will only be required in the final stages of a transaction, and then only if the product is highly complex and costly or if the technology is new and relatively unknown.

6.7 Savings and benefits

Cost savings can be made in all areas of business by use of a corporate intranet and the Internet. Examples of such savings are as follows:

- *Human resources* – information available to employees on line via a corporate intranet.
- *Training* – education and training on-line means employees do not have to leave the workplace to be trained.
- *Marketing* – major reductions in required paperwork, such as mail shots, manuals, product specification sheets and similar information, since this can be made available on-line for customers to read as and when required.
- *Workforce* – traditional sales personnel will no longer be required, as initial stages of sourcing and tendering will be via the Internet.
- *Telephone* – costs will be reduced, as communications flow between customers and suppliers using on-line technology. This is a much cheaper form of communication.

6.8 Techniques for building computer interfaces and environments

There are many different methods of constructing interfaces and environments. Many Web pages already contain high levels of interactivity with animated graphics and hyperlinks to other sites. Such links can be achieved by clicking on an icon, button or text. The format of the Web page can be in frames and data can be contained in tables. Many different types of software can be used to construct Web pages or as a standalone interactive front end. However, the boundaries between different types of software are blurring, for example many conventional programming packages can produce Web pages.

6.8.1 *Hypermedia*

This technique is generally considered to be a graphical programmed interface. It is also defined as 'object oriented programming'. For example objects such as buttons, pictures and three-dimensional forms can be modelled so that they interact with each other. Calculations can be made and files interrogated. It is basically a visual programming language. Examples of this language are Visual Basic, Visual C, Hypertext and Director by Macromedia. These languages can be transported to the Web.

6.8.2 *Multimedia*

The boundaries begin to blur as Web pages become the necessary communication medium within a company's network, internally as an intranet and externally as the

Internet. Various tools are available for Web authoring, some have already been mentioned. Here are a few more of the many that are available: Front Page, Dreamweaver, Netscape Composer. The choice depends upon the applications required and personal preference.

Web pages can be fully interactive with dynamic updating of pages from a database. This means that only two or three pages are required but the content can change within a template as data is transferred from a database. This technique uses Active Server Pages links. A good example of this is property advertising on the Web. Only a few pages are displayed, but details such as picture of the house, price, location and plan are dynamically updated from a database. If each page were unique and non-dynamic, thousands of individual pages would have to be generated: one dedicated page for each property. Using a template, the template remains the same and only the details within the template are changed from the database as required.

Other types of media include: audio, video, animated graphics (both two-dimensional and three-dimensional) and textual interface.

Examples of the broad range of hypermedia and multimedia databases are:

- encyclopaedias, dictionaries, manuals, handbooks and online documentation;
- learning and training systems, museum exhibits and interactive kiosks;
- ideas processing;
- collaborative work and computer-mediated communications;
- decisions-support systems and issue-based information systems;
- information management and information retrieval;
- software engineering;
- other applications, including simulation and modelling, law, WWW-user interfaces, design and organisation hypermedia.

It should now be clear how a manufacturing organisation can benefit by using these tools. It should also be equally clear that any manufacturing organisation not making use of these technological tools will be at such a severe disadvantage that its demise is guaranteed.

6.9 Online communities and business

The concept of online communities is now an established philosophy. For example:

- Companies of any size can develop internal and external networks of communication. They can open lines of communication between customers in a mutually beneficial way that enables a customer to hold a company more directly accountable for its services. Networked lines of communication using modern technology result in customers receiving more personal attention on urgent, important issues than is possible in a letter written to a nameless, faceless executive.
- A live company representative in regular communication via the Internet can form a personal bond with a potential customer, increasing the likelihood of closing a sale, something that could never be achieved through formal nameless and faceless written enquiries and written quotations.

- A customer's visualisation of different products, often using three-dimensional virtual images via the Web and then requesting help via the Internet, will help the customer choose a product.
- Collaboration with other companies over the Internet provides employees with advanced problem-solving tools and also saves company resources.
- Virtual meetings are very important, such as video conferencing (companies from around the world can meet in virtual board or seminar rooms where they can see and hear each other). They can also share data as they show off a new piece of software or conduct a seminar via white board software.
- The visual element of a shared application environment provides a vehicle for enhancing the tools available in the workplace, such as data visualisation and process simulation. For example, a team of engineers who design an aerospace engine by testing different parts online from different locations can benefit from a display of results that show cause and effect relationships in multiuser space.
- Extranets, where several intranets are linked together to form a fast reliable means of communication in multiuser environs, can help move a project toward completion. Salespeople from widely-dispersed car showrooms can use each other as a quick and easy resource when hunting down a customer request. Reporters from the different satellite bureaus of a newspaper group can work together on stories, from a virtual newsroom.
- Other ways in which online communities enhance company Web sites are by attracting and sustaining visitors with cutting-edge technology, promoting the company image through innovative branding campaigns, and generating revenue through electronic commerce and on-line sales.
- Electronic commerce in a shared environment is conducted in a familiar setting similar to that of the real world, and provides an entirely new and profitable commercial arena, where consumers are afforded interactivity and convenience. Unlike two-dimensional spaces, three-dimensional environments offer natural immersion where products can be viewed from all angles and, unlike the real world, buyers are not restricted by office hours, wrestling with parking or dealing with driving long distances to track down required products.
- In the areas of marketing and advertising, on-line communities provide an advanced way for companies to communicate their message and reach potential customers. These environments encourage an open dialogue between a company and its customer base, increasing loyalty to the company and its product line. Online communities can also serve as built-in, 24-hour focus groups, where company representatives can enter at any time to hear comments on the company or products performance.

    Customisation is the key to the marketing of products online, and online communities offer unique, new ways of allowing customers to receive personalised products and attention, including one-to-one relationships, marketing models rather than the 'hit and run' banner models.
- Many company Web sites now contain links to suppliers associated with their company. A supplier may pay a fee for this facility thus providing another line of income. This technique is known as Business to Business or B2B. For example, the Web site of a machine tool manufacturer could offer links to recommended suppliers of

tooling equipment, in-process-measuring equipment, workholding equipment and automated and compatible load/unload equipment.

However, there is a downside to the overuse of and over-reliance on electronic communications as exemplified by many 'call-centres'. Here frustrated customers, after waiting for an interminable length of time, are confronted with a request for a large amount of 'security' information that must be provided before their request for a simple item of information can be processed. Often the data required is not immediately to hand and the call has to be terminated, the data has to be found and the whole process has to be started again. Equally annoying are automated telephone answering services that offer a menu of choices that never seem to include the information that is required. Again, there is no excuse for supervisory and management staff to 'hide behind' their call-centre staff so that it becomes impossible to speak to a person in authority. Any technologically-advanced communications system – particularly when dealing with the general public – must be user friendly if it is not to become self defeating in its aims of greater efficiency. It must genuinely exist to provide improved customer service and not just be a vehicle for cost-cutting parsimony.

6.10 Management of manufacture

The earliest manufacturing organisations had owner-manager supervision and many small businesses are operated successfully in a similar manner. One of the first decisions that has to be made in any organisation is which of the many duties must the owner-manager undertake personally. Even in a 'one-man' business materials have to be purchased, products manufactured, invoices prepared, jobs costed and tenders submitted, in addition to the raising of capital, and tax considerations. The problems become magnified as the organisation expands and labour has to be hired and paid for, and employment legislation has to be complied with. The high failure rate of small businesses gives testimony to the fact that many persons set up in business on the basis of their technical prowess and craft skills but without the necessary organisational skills or without giving sufficient thought to the problems of being one's 'own boss'. Managing an enterprise either alone or, as is more usual, as part of a team requires training and skill in just the same way as does operating a machine tool. Small- to medium-sized companies engaged in engineering activities did, and still do, appoint supervisors who are more skilled than the most-skilled workers under their control. Such supervisors also have responsibility for production planning, control of quantity and quality, and the dispatch of finished goods. Increasingly, however, even small and medium enterprises (SMEs) need to adopt the principles and practices of scientific management in order to survive.

The scientific approach to management involves the application of the following principles:

- *Forecasting and planning*: This requires an assessment of future needs and decision making for future action.
- *Organisation*: The division of labour, the allocation of duties, and the lines of authority and responsibility.
- *Command*: The issuing of instructions to ensure that decisions are activated.

Fig. 6.1 *Structure of relationships*

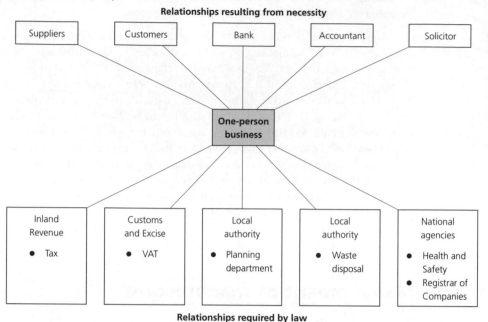

- *Control*: The setting of standards, the comparison of physical events against the set standards, and the taking of any necessary corrective action.
- *Coordination*: The unification of effort to ensure that all activities of the business are pursuing the same objectives.
- *Communication*: The transfer of information between different people and/or sections of the organisation, in particular between management and the workforce. Also communication between the company and its customers, suppliers and allied services, as shown in Fig. 6.1.
- *Motivation*: This is the driving force behind all actions. Psychological considerations make it important to recognise the motivation behind the customer buying the goods or services supplied by the organisation and the motivation of the people working in the organisation.

In addition to the above principles of management is the introduction of *behavioural science*. This assists in an understanding of human behaviour, for example, job satisfaction, attitude to work and putting the right person in the right job. Such skills are essential to all managers because 'management is usually about people'.

6.10.1 *'World class' manufacturing*

It is essential for all companies to aim to become 'world class' if they are to succeed and survive in the global economy. The principles involved apply to all companies, large and small. Even small firms can no longer afford to be parochial in their market outlook if they wish to expand and prosper. Let us see how 'world class' can be defined.

A 'world class' organisation is one that is continuously more profitable than its competitors. This can only be achieved by its being more successful than its competitors in satisfying customer needs. World class can therefore best be defined in terms of customer satisfaction, for example a company that provides high-quality products or services (Q), at less cost (C), and with shorter lead times that enable it to meet its delivery targets (D), should be world class. Since all employees in the organisation may also be viewed as internal customers from the 'total company' point of view, a high level of worker safety (S) and morale (M) can be added to the definition.

A scoreboard should be developed to measure progress in relation to each of these objectives. The following are examples of QCDSM measurement for a scoreboard:

- *Quality*: Rework; number of customer complaints; defects.
- *Cost*: Productivity; overtime; floor space.
- *Delivery*: Conformance to schedule; lead-time; volume of production.
- *Safety*: Number of accidents; number of safety related suggestions.
- *Morale*: Absentee rate; turnover rate; number of suggestions.

Becoming 'world class' is a long-term process that requires continuous improvement and innovation to better the current level of achievement. The following action agenda can be followed to implement 'world class' principles in the manufacturing environment:

- *Understanding the basics*. All employees must understand 'world class' principles. This can be achieved through study sessions and tours of successful 'world class' plants.
- *Reducing change-over times*. Use external instead of internal set-up. For example the tooling is palletised at the side of a press so that the tools currently in use can be rolled out and the replacement tools, externally set, can be rolled in ready for immediate use.
- *Improving plant layout*. Minimise transport of materials by changing to U-shaped or parallel lines and cellular manufacturing. Arrange the workplace to eliminate search time. The adoption of JIT methods of working minimises 'work in progress' having to be stored adjacent to the processing site, using up valuable floor space, or having to be brought from stores, which involves additional labour and transport facilities.
- *Increasing worker training*. Multiskilling personnel leads to an increase in their flexibility.
- *Increasing worker responsibility*. For example, source inspection where each piece received from the previous process is checked by the operator and defects are reported immediately:
  - (a) *Problem-solving*. Line personnel must not only be primarily responsible for basic problem-solving but must be adequately trained and have access to the necessary information needed to solve problems as they arise.
  - (b) *Multiple machines/processes*. Make a worker responsible for more than one machine or process.
  - (c) *Data collection*. Personnel must record and retain data on production, quality and problems at the workplace.
- *Improving the product and process*. Make it easy to manufacture without error.

- *Reducing inventory.* Cut inventory levels to a minimum with the goal of zero inventory. Again JIT helps to achieve this goal.
- *Reducing lot sizes.* Reduce the volume of 'work in progress' and flow times by reducing lot sizes.
- *Selective introduction of new equipment.* Maintain and improve existing equipment and human working practices before thinking of new equipment:
  - (a) automate gradually, when process variability cannot otherwise be reduced;
  - (b) use machines with abnormality-detection capability.
- *Levelled/mixed production.* Produce multiple products on the same line and provide upstream processes with balanced loads. For example variants of trim, colour and accessories (optional extras) for automobiles, to suit individual customer requirements, are now usually made on the same production line.
- *Supplier cooperation.*
  - (a) Introduce suppliers to 'world class' principles and help them to put the principles into practice.
  - (b) Cut number of suppliers down to a few reliable ones. By increasing their batch sizes, costs can be reduced, through economy of scale, leading to increased automation.
  - (c) Cut the number of parts variants by standardisation. Use British Standard (BS) components wherever possible to ensure availability, reduced cost and approved quality.
- *Introducing a kanban system.* Use a simple *kanban* system to move from a push to a pull scheduling system. Kanban (introduced by the Toyota car company and meaning *card*), in its simplest form a just-in-time system, is a card system. In this system assembly schedules are derived from a master plan, which is drawn up to meet a specific 'planning horizon'. The planning horizon is the time period that has previously been agreed as realistic and trustworthy. The daily production requirements can be set by using a master plan and the agreed timescale. For example a card authorising the dispatch of an agreed quantity (*bin*) of parts to the assembly line would be issued to the stores. The dispatch requirements would be noted on the card and it would then be issued to the production department, so that another set of freshly manufactured parts could be 'pulled' into the stores. The process could be computerised rather than relying on the physical movement of bin cards but the principle remains the same.
- *Improving customer relations.*
  - (a) Get to know the customer's needs.
  - (b) Increase make on deliver frequency for each required item.

6.11 Materials requirement planning

Most organisations 'buy in' materials in one of the following ways:

- In the raw state for processing into a finished product; in a part finished state (e.g. castings or forgings) for processing into a finished product.
- As standard components (e.g. nuts and bolts) ready for assembly.
- As a mixture of one or more of these foundation materials.

The cost of foundation materials can form a large part of the final cost of the finished product. If the elapsed time between the delivery of the foundation materials and the sale of the finished product is small, then the cost of holding stock is low. However, if it is high, then the cost of holding stock is significant. Because there is no return on the working capital invested in stocks of foundation material, this can adversely affect the profitability of the company. At worst it can adversely affect the 'cash flow' of the company because funds tied up in the material represent 'dead money' that cannot be used for more profitable purposes. Furthermore, if bank borrowing is necessary for the purchase of the material, the interest charges can turn a paper profit into an actual loss.

6.11.1 *Stock control*

The objective of materials control is to maintain a supply of foundation materials at the *lowest possible cost*. A term that is frequently used in *stock control* is 'lead-time'. This is the time taken from the initiation of an order to the delivery of the material into the company's stores. If the foundation material is a stock item at the supplier, the lead-time will be short. However, if the material has to be manufactured by the supplier, the lead-time can become very significant. Thus the problem is that of raising orders at the correct time, both from outside suppliers and from internal production departments. The term MRP has been introduced earlier in this chapter and stands, variously, for *material requirements planning* (MRP 1) and *manufacturing requirements planning* (MRP 2).

Material requirements planning has been in industry for a very long time but with the advances made in recent years in computer technology and software, a computer-based system for material requirements planning has been developed. Before examining MRP 1 in detail, let us review the more common methods of organising stock-control in a company.

The term 'stock-control' describes a system of purchasing in which the level of stocks being held in a store is used to regulate the raising of buying or production orders. Its function is to ensure that an adequate supply of foundation materials is available at the correct time and in the correct quantities to provide continuous production. Items can be bought-in via the purchasing office or manufactured 'in-house'. For the reasons previously mentioned, stocks should be kept to the minimum necessary to maintain production. The amount of foundation material held in stock is governed by operational needs, lead-time, cost of storage, deterioration of materials while in stock and the cash-flow position of the company – stock ties up working capital.

Stocks should only be held if:

- delivery cannot be exactly matched to supply;
- delivery is uncertain;
- substantial price reductions provide a cost advantage;
- bulk buying attracts discounts that offset the cost of storage;
- they provide a buffer of finished products providing a service to customers;
- in-house production is prone to operational risks (process breakdowns).

6.12 Stock control systems

The basic feature of any system of stock control is that the *stock level initiates any new orders*. When the stock level falls to some predetermined level – 're-order level' or 'order-point' – then a new order is issued for a further supply. Figure 6.2 shows the basic features of a stock-control ordering system based upon variation of stock over a time period. Note that:

- the *order point* (OP) is the level of stock at which a new order is issued;
- the *order quantity* (OQ) is the quantity ordered to restore stocks to the agreed level – also known as the 'batch quantity' or 'batch size';
- the lead-time is the time required either to buy in or to manufacture, in house, a replacement batch;
- if the batch quantity is known, then the lead-time can be calculated.
- the order point can be determined by setting off the lead-time and batch quantity.

Fig. 6.2 *Simple stock control ordering system*

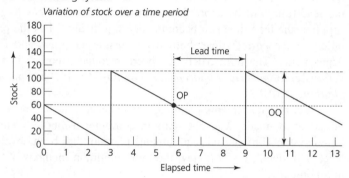

The steps required to set up an elementary stock control system are as follows:

1. Choose the order quantity.
2. Estimate the lead-time to obtain a replacement batch of the required size.
3. Calculate the number of parts that will last, at the normal consumption rate, during the lead-time, and thus find the order point.
4. Maintain a permanent record of the stocks of each item.
5. Arrange for a new order to be issued each time that the stock level falls to the order point.

6.12.1 *Buffer stock*

The simple system shown in Fig. 6.2 is impracticable because of the imprecision of actual production completion times. If a delay in production occurs and the lead-time increases or there is an increase in demand, then stocks would run out causing an interruption in production at some stage. An improved model is shown in Fig. 6.3. The buffer stock represents an insurance against the risk of the exhaustion of stock. Its value is obtained

Fig. 6.3 *Minimum and maximum stock levels*

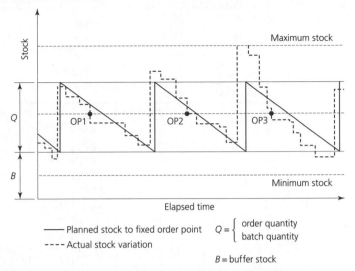

by balancing the cost of a shortage, causing interrupted production, against the working capital locked up in holding additional stock.

6.12.2 *Stock level*

The *minimum stock* is a control level that reduces the risk of exhausting stocks, as shown in Fig. 6.3. When stocks reach the *minimum stock level*, the fact is reported so that an investigation can take place and corrective action can be taken. Buffer stock (*B*) and minimum stock need not be the same amount because the former is a fixed amount of stock maintained as an insurance against high demand or delays in delivery, while the latter is a control level.

The maximum stock is also a control level and indicates when stock levels are unacceptably high in economic terms (see Fig. 6.3). When this point is reached, production of that stock must be reduced or even ceased until stocks have reached a more manageable level. The maximum stock level is set so that it is in excess of buffer stock by an amount sufficient to allow for small variations in the production plan. Only when something is wrong in the system are the maximum and minimum stock levels normally reached. A typical stock control record card is shown in Fig. 6.4.

6.12.3 *Batch quantity*

Although it was originally thought possible to calculate the *economic batch quantity* (EBQ) using mathematical formulae, EBQ is now commonly selected for each case on merit and previous experience. That is:

Batch quantity = (total requirement per year)/(batch frequency)

This simple calculation avoids batch quantities being set to suit production convenience without regard to stock costing.

Fig. 6.4 *Stock control record card*

Batch quantity:	100			Description:			
Order point:	56						
Buffer stock:	40			Bearing Block			
Maximum stock:	160			Part No.: 3890 A			
Minimum stock:	30						
Date	*In*	*Out*	*Balance*	*Date*	*In*	*Out*	*Balance*
3.6.91	98	–	98	2.9.91	–	12	36
10.6.91	–	12	86	9.9.91	_	2	34
17.6.91	–	16	70	16.9.91	104	–	138
24.6.91	–	14	56	23.9.91	–	12	126
1.7.91	–	14	42	30.9.91			
8.7.91	–	4	38				
15.7.91	–	14	24				
22.7.91	102	–	126				
29.7.91	–	16	110				
5.8.91	–	16	94				
12.8.91	–	14	80				
19.8.91	–	16	64				
26.8.91	–	16	48				

Significant Dates
24.6.91 – Re-order
15.7.91 – Minimum breached; investigate
26.8.91 – Re-order

6.13 Materials requirement planning (demand patterns)

Let us consider an everyday item such as a bicycle. Figure 6.5 shows the relationship between the component parts of the bicycle. The completed assembly is at the highest level and is classified as level 0. The items making up the constituent parts are classified as the lower levels 1, 2, 3 and 4. Therefore, if the items classified at these lower levels are not being sold as spare parts, they can be classified as *dependent* items. That is the demand is solely dependent upon the number of complete bicycles assembled.

The completed bicycle is an end in itself and it is not used in, or as part of, a higher-level item. It is, in fact, the finished product that the customer purchases from the local cycle shop. The volume of production is based upon sales forecasts for that particular model and is *independent* of the demand for any constituent parts. Thus the demand for a company's products can be classified into two distinct categories:

Fig. 6.5 *Materials requirement planning*

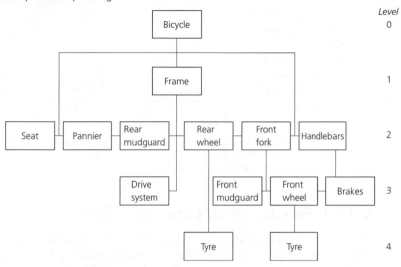

- *Dependent demand* generated from some higher-level item. It is the demand for a component that is dependent upon the number of assemblies being manufactured. As such, it can be accurately calculated and inserted into the *master schedule*.
- *Independent demand* generated by the market for the finished product. The magnitude of the independent demand is determined from firm customer orders or by market forecasting. For MRP 1 purposes it is independent of other items. It is not used in scheduling any further assemblies.

A stock-control system could keep all the parts shown in Fig. 6.5 in stock using either order-point methods or sales forecasts for each item. To do this each separate item has to be treated as being independent. The MRP 1 software can, however, calculate the amount of stock of dependent items required to meet the projected demand for the independent higher level item that, in this example, is a bicycle.

It is claimed that MRP 1 is superior to the standard methods of stock control previously discussed in this chapter. This is because, while the demand for the independent items (level 0) cannot be accurately predicted – and is therefore unsuitable for stock control methods – lower level items (1, 2, 3, 4) are dependent on a higher level item and can thus be calculated with accuracy. That is, although it is only possible to forecast next year's sales of complete bicycles approximately, it is possible to predict that exactly as many tyres will be required as there are wheels.

Furthermore, stock control methods assume that usage is at a gradual and continuous rate. In practice, however, this is not the case. The call on parts will occur intermittently, with the quantity depending upon the batch size of the final product. This often results in excessive stocks being held. Materials requirement planning assists in the planning of orders so that subsequent deliveries arrive just prior to manufacturing requirements, as shown in Fig. 6.6.

Fig. 6.6 *Comparison: stock control versus MRP 1*

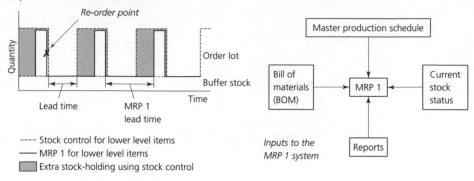

6.13.1 *Master production schedule*

This is the schedule or list of the final products (*independent* or *higher level* items) that have to be produced. It shows these products against a time or period base so that the delivery requirements can be assessed. It is, in fact, the definitive production plan that has been derived from firm orders and market forecasts. In order to minimise the number of components specified, care has to be exercised in deciding how many of the basic level 0 assemblies are to be completed. The total lengths of the time periods (the *planning horizon*) should be sufficient to meet the longest lead-times.

6.13.2 *Bill of materials*

The bill of materials (BOM) contains all the information regarding the build up or structure of the final product. This will originate from the design office and will contain such things as the assembly of components and their quantities, together with the order of assembly. The structure and presentation of a typical BOM is shown in Fig. 6.7. The number of parts should also be shown at each level.

6.13.3 *Current stock status*

This is the inventory or latest position of the stock in hand, pending deliveries, planned orders, lead-time and order-batch, quantity policy. All components, whether in stores, off-site or in progress, need to be accurately recorded in the inventory record file.

6.14 MRP 1 in action

Upon receiving the input data, the system has to compute the size and timing of the orders for the components required to complete the designated number of the final product (the independent or highest level item). It performs this operation by completing a level-by-level calculation of the requirements for each component. Since the latest stock information is available to the system, these calculations will be the net value of the orders to be raised.

Fig. 6.7 *Structure for bill of materials: S = sub-assemblies; C = components*

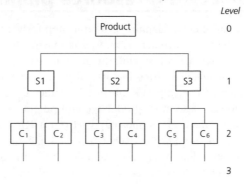

Fig. 6.8 *Standard planning sheet for one of the outputs of MRP 1*

	1	2	3	4	5	6	7	8
Gross requirements				140				
Scheduled receipts			50					
On-hand	100		150					
Net requirements				100				
Planned order releases				100				

requirements = net requirements + stock currently available

where

stock currently available = current stockholding + expected deliveries

Inherent in the calculations is the factor for lead-time, i.e. ordering and manufacturing times. The system must determine the start date for the various sub-assemblies required by taking into account the lead-times, i.e. it offsets these lead-times to arrive at the start dates. A complication that the system has to overcome is the use of some items that may be common to several sub-assemblies. The requirement for these common-use items must then be collected by the system to provide one single total for each of these items. Figure 6.8 shows a standard planning sheet for one of the outputs of MRP 1. Other reports required in addition to the order release notices are:

- re-scheduling notices indicating changes;
- cancellation notices where changes to the master schedule have been made;
- future planned-order, release dates;
- performance indicators of such things as costs, stock usage and comparison of lead-times.

A number of advantages are claimed for MRP 1 software. These include:

- a reduction in stocks held in such things as raw materials, work in progress and 'bought-in' components;
- better service to the customer (delivery dates are met);
- reductions in costs and improved cash flow;
- better productivity and responses to changes in demand.

6.15 Manufacturing resource planning

Materials requirements planning is a large step forward, compared with manual systems, in the procurement of materials and parts inasmuch as it provides a statement of what exactly is required and when. Unfortunately, it takes no account of the production capacity of a particular manufacturing facility. Most factories have a production output that can only be changed over a period of time by either improved productivity or the injection of capital to purchase more productive equipment.

Clearly, by themselves, the outputs from MRP 1 have limited value to the *master production schedule* (MPS) except to detail the materials requirements. Without the information regarding production capacity, it is difficult to match MRP 1 to actual production schedules. Remember that *capacity planning* is concerned with matching the production requirements to available resources (human resources and plant). It would be impractical to have a master production schedule that exceeded the capacity of the plant. This can lead to decisions to extend the plant capacity by increasing the labour force, shift working or sub-contracting, when what is required is more sensible scheduling.

The shortcomings of MRP 1 led to the development of MRP 2 that brought into account the whole of a company's resources, including finance. Thus the initials MRP 2 stand for *manufacturing resource planning*. This is a quite different, and more difficult, philosophy than the original *materials requirements planning* when the many variables that can affect shop floor production are considered. For example, machines may breakdown, staff may become ill or leave, and rejects may have to be reworked. In addition, MRP 2 takes into account the financial side of the company's business by incorporating the *business plan*. Manufacturing resource planning is used to produce reports detailing materials planning together with the detailed capacity plans. This enables control to be effected at both the shop floor and the procurement levels. Note that MRP 2 does not replace MRP 1 but is complementary to it and *the two systems are used together*.

As in all computer systems, the information produced is only as good as the information fed into it. Therefore the system is dependent upon the feedback of information relating to those things under its control, i.e. such data as the state of the manufacturing process and the position of orders. Manufacturing resource planning contains enough information, and the power to organise it, to be able to run the whole factory. Figure 6.9 shows a schematic layout of a production control system incorporating MRP 1 and MRP 2 organised around the master production schedule.

Manufacture resource planning is a management technique for highlighting a company's objectives and breaking them down into detailed areas of responsibility for their implementation. It involves the whole plant and not just the materials provision. The key to MRP 2 lies in the master production schedule (also called the *mission statement*), that is a statement of what the company is planning to manufacture and, because it is the master, all other schedules should be derived from it. The master schedule should *not* be a statement of what the company would like to produce, as its own derivation lies in the company objectives. It cannot in itself reduce the company's lead-times, but it will provide the incentive for the plant personnel to take action if required. It is important, as in all planning, to monitor adherence to the master production schedule. Manufacture resource planning will not stop over-ambitious planning but it will show up the consequences.

Fig. 6.9 *Schematic layout of MRP 1 and MRP 2 organised around the master production schedule*

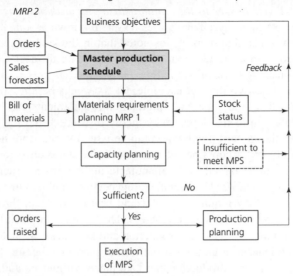

6.16 Just-in-time

The phrase 'just-in-time' (JIT) originated in Japan with the production plant of Toyota as the most frequently quoted application. This led to the incorrect conclusion that since Toyota is involved with mass (repetitive) production of cars, then JIT only applies to this mode of manufacture. In fact JIT applies to any mode of manufacture. In the USA the term 'zero inventories' is used but the philosophy is exactly the same as JIT. Perhaps the most apt definition of JIT belongs to R. Schonberger (*Japanese Manufacturing Techniques*, Free Press, New York, 1982),

> Produce and deliver finished goods *just-in-time* to be sold, sub-assemblies *just-in-time* to be assembled into finished goods, fabricated parts *just-in-time* to go into sub-assemblies, and purchased materials *just-in-time* to be transformed into fabricated parts.

A second approach is propounded by D. Potts (*Engineering Computers*, September 1986), 'A philosophy directed towards the elimination of waste, where waste is anything that adds to the cost but not the value of the product.'

A third approach that appears to combine the other two is put forward by C. Voss (*Just-in-Time Manufacture*, IFS, London, 1987), 'An approach that ensures that the right quantities are purchased and made at the right time and in the right quantity and there is no waste.'

So there we have it: JIT philosophy is not only about minimum stocks being available at the point of manufacture. It is also about reducing costs by eliminating those things that add nothing to the value of the product, such as avoiding interest charges on working capital tied up in servicing unnecessarily large stocks. At the same time, it is necessary that the availability and quality of the product is maintained. However, any manufacturing system should have the foregoing objectives, so it is in the application that *JIT gets nearer to the manufacturing ideal than many traditional systems.*

As stated earlier, Toyota uses a system known as *kanban*, so, in this form, their JIT system is a card system. It was noted that in the kanban system assembly schedules are derived from a master plan, which is drawn up to meet a specific 'planning horizon'. The 'planning horizon' is the time period that has previously been agreed as being realistic and trustworthy, for example, let six weeks be appropriate. Using the master plan and the time-scale, the daily production requirements can be set.

Referring to the dependency theory of MRP 1 and the bicycle example considered earlier, then a card (*kanban*) would be issued to the store holding the parts for the final assembly and a container of parts would be dispatched to the assembly point. A card would then be issued to the manufacturing centres for them to make replacements for the final parts store. The manufacturing centres would, in turn, issue a card for any replacements that they require. Cards would be passed down through the system to raw materials purchase and sub-contractors. Thus, ideally, at each level the replacements arrive 'just-in-time' to maintain production and no stocks are held.

The issue of a card from the master plan or schedule 'pulls' the work through the system, since no production can take place without its authorisation. Figure 6.10 illustrates this point. In such a 'pull-through' system in which everything is dependent upon the arrival of an authorisation card, there must be times when no production takes place. This is because of the over-riding principle that nothing is produced until it is required except, of course, for the minimum (small) inventories kept in the appropriate stores.

The impression might be given that the labour force sits around doing nothing until a card arrives. This is not the case. The major advantage gained from any inactivity is that the reason for that stoppage can be immediately investigated. Everyone – management and production workers – can be engaged upon tackling the problem. In the meantime,

Fig. 6.10 *Flow diagram for a simplified kanban system*

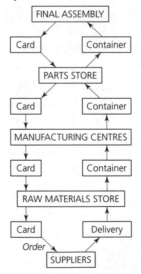

those members of the production team not engaged directly in solving the problem can be engaged in quality circle work, training or maintenance activities.

The Toyota system is just one example of the philosophy of JIT because, at the end of the day, that is what JIT is – simply a philosophy. It covers a range of production control techniques that have the objectives of eliminating every facet of the business that fails to make a contribution. It focuses minds on achieving production that is not only 'just-in-time' but which produces 'just enough' with no surplus and costly excess stocks. It streamlines the production process by removing the cushion of safety stocks, excess stocks and/or inflated lead-times that can cover up many real problems in a company. It concentrates the minds of the entire workforce on the business of the company. By making every action important, there has to be involvement and commitment from everyone for the successful implementation of JIT. A *team approach to problem-solving* is inherent in JIT.

Without a JIT philosophy, 'waste' can be overlooked in a company because it is not recognised and because of a 'we have always done it this way' mentality. When waste is considered to be any activity failing to contribute to the addition of value to the product, it can cover such things as the production of scrap and reworking, excessive transportation by poor routing and poor plant layout, overproduction to compensate for scrap, machine breakdowns and absenteeism. All of these things can too easily become part of the production scene by the acceptance of their inevitability. The analysis of management and working practices that are necessary when introducing the JIT philosophy immediately shows up these faults in the system.

Note that each element in a company is dependent on all other elements and therefore JIT can become a very vulnerable system. Employees and suppliers are all part of the system so that any conflict will halt production very quickly. In a large organisation, for example, an industrial dispute in one link in the supply chain can quickly bring the rest of the chain to a halt.

6.17 Just-in-time and MRP 2

The *kanban* (card) system just described is only one way of applying the JIT philosophy. As production becomes more varied and involved, then the amount of data required to operate the system increases significantly and *computer support is required*. To this end MRP 2, which was introduced in Section 6.15, is a computer-based system that has the objective of unifying all the associated functions in a company from initial planning through to delivery of the finished products. It is applicable to any form of company irrespective of size and diversity of production. It breaks down the company's *business plan* into detailed operational tasks for each part of the organisation. The effective planning element removes any confusion that could exist by concentrating the minds of the workforce – management and productive labour – on what has to be achieved.

Because JIT has the major objective of eliminating waste in all its forms and confusion leads to wasteful excesses, the adoption of MRP 2 promotes the concept of JIT. The key to successful implementation is correct planning and scheduling. Both have the main aim of producing only what is required and when it is required and are therefore compatible. Thus, the benefits of JIT can be summarised as:

- better quality by closer control;
- reduction of lead-times;
- reduction of stocks and levels of work in progress;
- more involvement of the workforce leading to greater job satisfaction;
- increased efficiency by the process of continually challenging the existence of everything connected with the production in a never-ending search for greater efficiency.

6.18 Computer integrated manufacture

The above techniques are all integrated into a seamless system where data is shared between all functions within any department. These systems contribute towards the philosophy of computer integrated manufacture (CIM), as shown in Fig. 6.11.

Fig. 6.11 *Computer-integrated manufacture (CIM)*

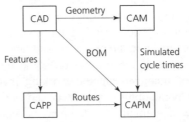

6.18.1 *Computer aided process planning*

Computer aided process planning (CAPP) reduces the amount of manual clerical work involved in planning the processes required to manufacture a component, sub-assembly or assembly. This is achieved by integrating or sharing data with other computer systems, such as computer aided design (CAD). Geometrical data or features from such systems can be used to decide which processes will be used. Other computer systems, such as manufacturing databases, can also be used to automate the planning process. For example databases can contain previous process data on components that share similar family features, i.e. a family of components. Coding systems can be used to identify components and group technology can be used to group components into features-based families. The route each component takes as it passes through each process towards its finished state is then passed to the *computer aided production management* (CAPM) system.

6.18.2 *Computer aided production management*

Computer aided production management is again another layer of integration with other systems. For example *bill of materials* data, automatically generated from CAD systems, can be passed to CAPM systems along with routing data for each component required. The purpose of CAPM is to schedule and prioritise the sequence of events within a manufacturing system. Again, data from tool paths simulated in computer aided manufacturing systems can be used to derive cycle times. This enables the most efficient schedule to be developed so that lines are balanced, capacity is planned, people and machines are fully utilised and down time is minimised.

6.18.3 *Product data management systems*

Product data management systems (PDMSs) are now used to manage data within a company. Such systems as an *intranet*, that is an internal web system for a company, enable data to be shared at all levels. Many companies are international corporations and data needs to be shared among the many plants of such a company on a global basis. For example some aircraft companies share design data across the Atlantic using 'intranet' systems. Other benefits, such as developing designs based on previous process information, are useful in a 'design for manufacture' context. Many companies are very close to becoming *closed-loop* systems, i.e. *shop-floor, data-capture systems feedback data*, which can be used to monitor progress.

6.19 Quality in manufacture

As stated at the end of Chapter 5, quality control is a 'systems led' management function and as such is readily computerised. We will now consider some of the basic principles of quality control without concerning ourselves with the computer systems required for data gathering and processing and recording, as these will vary according to the manufacturing process involved. Not only does the keeping of computer records substantially reduce the storage space required by hand-written paper records but also, if statistical methods are employed, the calculations involved can be speeded-up by the use of computer templates and human error can be removed.

Figure 6.12 shows the importance of the customer in manufacture. The concept of supplier and customer exists within an organisation to provide a *quality chain* between

Fig. 6.12 *The linkage between customer expectations, product design and manufacture*

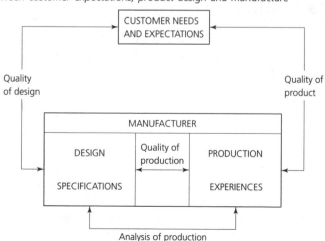

Fig. 6.13 *Quality chain*

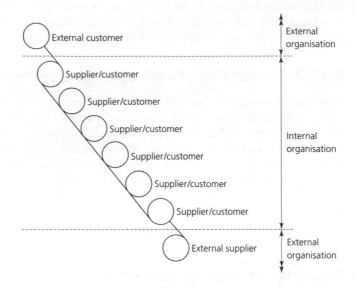

the external supplier to the company and the delivery of goods to the customer. Each department is the customer of the departments supplying it with goods and services and, in turn, each department becomes the supplier of those departments making use of its goods and services. Similarly, within each department, there is a supplier/customer relationship between the individual members of the personnel of the department. Typists are suppliers of documentation to their managers who, as the customers, have the right to expect a high-quality, error-free typing service.

Throughout and beyond all organisations, be they manufacturing concerns, banks, retail stores, universities or hotels, there exists a series of quality chains, as shown in Fig. 6.13. This chain may be broken if one person or one piece of equipment does not meet the requirements of the customer, internal or external.

6.20 Reliability

This is the ability to provide 'fitness for purpose' over a period of time. Obviously any component will fail at some stage of its life, although the time span may be such that the component seems to be virtually everlasting. Failure is considered to be the point at which a product or service no longer meets its fitness for purpose. Increasingly, reliability is a major factor when a customer makes a purchasing decision.

Therefore there is an increasing need to design reliability into a product from the start. The testing of a new design for its reliability is difficult due to the time involved as well as the widely-varying environments and conditions under which the product has to operate during its working life. There are many tests specifically designed with re-liability in mind. For example, in the aircraft industry, there are rigs set up to test the fatigue resistance of wing assemblies. Hydraulically actuated pistons are positioned along the underside of a wing and are programmed to move up and down to simulate the forces acting on the wing during flight conditions. By this method, many thousands of flying hours can be compressed into days while the assemblies are safely tested to destruction. Again, rolling roads and specially designed test tracks are widely used in the automobile industry. In addition, prototype vehicles are test driven under varying climatic conditions over very large distances throughout the world to test their reliability before entering large-scale production and being sold to the motoring public.

From the outset, a designer can take some practical steps to prevent premature failure. For example:

- the use of only components with a proven reliability in assemblies. This can prove to be difficult when a designer is being innovative and is introducing new materials and manufacturing methods;
- the use of uncomplicated designs incorporating as many previously proven features as possible;
- the use, within the cost parameters, of back-up systems and duplicated components where total failure cannot be tolerated, as in the controls of automobiles or aircraft. Duplication decreases the probability of total failure by the square of the chance of a single component failing. Triplication in parallel of a number of components decreases the probability of failure by the cube of the chance of a single component failing. Wherever possible components should be designed to be fail-safe. That is, failure of the component shuts down the system of which it is a part or a back-up system is automatically brought into operation. The pilot light on a modern gas appliance must be so designed that, should it fail, the gas supply to the appliance is automatically cut off;
- only tried and tested methods of manufacture should be employed until new methods have been proved to be equally reliable or better. This may sound dull to a thrusting young designer, but the cost of failure can be prohibitive both financially and in the loss of human life. This is particularly true in the present day climate of compensation culture.

6.21 Specification and conformance

Accepting that the customers' needs are paramount when considering the definition of quality as 'fitness for purpose', it becomes necessary to establish at the outset exactly what those needs are. Figure 6.12 shows how the customer has to be involved at every stage from design through to delivery. In the production of goods or the delivery of services, there are always two separate but interconnected factors relating to quality: quality in design and quality of the conformance with the design. Let us now look at these two factors in some detail.

6.21.1 *Quality in design*

The first stage in the development of a product or a service is 'customer needs'. This is the province of the marketing department working in conjunction with the design engineers. This first stage is the most difficult and yet the most crucial as it sets the scene for the whole process of manufacture in the case of goods, and delivery in the case of services. The term 'quality of design' is a measure of how the design meets with customer's requirements, i.e. the stated purpose of the *design brief* and the *design specification* derived from it.

The translation of customer needs, as set out in the initial design, into manufactured goods or delivered services is achieved by the preparation of specifications. These specifications can best be described as detailed statements of the requirements with which the goods or services must conform. Correct specifications control the purchase of goods or services as well as directing the manufacturing processes. Specifications can be expressed in terms of such things as mechanical properties of materials, tolerances on machine parts and surface finish requirements, all of which are easily recognisable in engineering terms. Specifications may also refer to more abstract concepts such as the *aesthetics* of design. Furthermore, specifications can also apply to the internal supplier/customer relationship. For example, a designer may request the company's legal department to draw up a specification for a product such that it does not infringe any patent, copyright or legislation, such as that concerned with product liability.

The translation of customer needs (the *demand specification*) into a *design specification* and then into a feasible *manufacturing specification* is very time consuming. This is because each demand need has to be broken down into sub-needs that are then issued as either purchasing orders or manufacturing requirements. The design specification must contain information concerning the minimum functional, reliability and safety requirements together with the information necessary to achieve minimum cost factors. Thus the concept of minimum functional requirements is closely related to the achievement of minimum cost factors. Any designer must realise that demanding higher specifications, such as greater precision, leads to an increase in manufacturing costs. Figure 6.14 shows the influence of the design specifications on the value added to a product. Increased precision often results in increased production costs, leading to an initial rise in the selling price followed by decreasing sales and returns, as indicated by the added value curve. To demonstrate that the 'quality of design' meets customer demand, the design engineers can test materials and components, and build and test prototypes. Customer involvement at this stage is essential in order to secure type acceptance before quantity production commences.

Fig. 6.14 *The influence of added value on design specification (increased precision = an initial rise in the selling price followed by decreasing returns)*

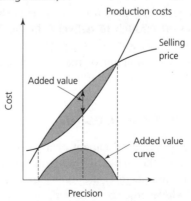

6.21.2 *Quality of conformance to design*

The ability to manufacture goods or deliver a service to agreed specifications at the point of acceptance is referred to as the *quality of conformance*. Many companies now employ conformance engineers whose function is to ensure that goods and services match the design specifications and hence the demand specifications.

Quality cannot be inspected into goods or services. The traditional concept of quality control consisted of inspecting samples of components to ensure that they had been made 'according to the drawings' and rejecting any components that were out of tolerance. Unfortunately this concept that quality is solely the province of the inspection department still persists in many companies.

If the philosophy of customer satisfaction and customer demand led design is adopted from the start, this philosophy must be followed throughout the whole system from the basic idea to the finished product or service. This is achieved by regular conformance checks during production to ensure that the specifications are being achieved. *Thus, conformance testing is about checking the quality of the whole process and **not** just about inspection of the finished product.* This avoids the waste of making rejects and the costs of reclamation and rectification. A high level of end-inspection indicates that a company is trying to 'inspect-in' quality. The retrieval and analysis of production data plays a very significant part in achieving specified levels of quality, so that when the product or service reaches the customer and *fully satisfies* all the specified requirements, then *quality of conformance* will have been achieved. Figure 6.15 shows the relationships between inputs and outcomes of quality conformance.

Fig. 6.15 *Quality inputs and outcomes*

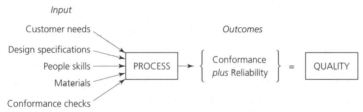

6.22 Accreditation of conformance

In demanding and competitive markets, good product design and efficient manufacturing must be underpinned by properly authenticated measurement and testing. For a company to have absolute confidence when issuing a certificate of testing, its own inspection instruments must, themselves, have been certificated by an independent accreditation agency. One such agency is the National Measurement Accreditation Service (NAMAS).

NAMAS is a service offered by the National Physical Laboratory (NPL), one of the research establishments of the UK's Department of Trade and Industry. NAMAS assesses, accredits and monitors calibration and testing laboratories. Subject to its

Fig. 6.16 *National measurement system*

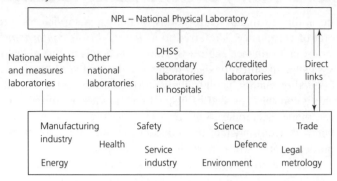

stringent requirements, these laboratories are then authorised to issue formal certificates and reports for specific types of measurements and tests. NAMAS accreditation is voluntary and open to any UK laboratory performing objective calibrations or tests. This includes independent commercial calibration laboratories and test houses, and also laboratories that form part of larger organisations, such as a manufacturing company, educational establishment or Government department. It includes cases where the laboratory concerned provides a service solely for its parent body.

Every industrialised country requires a sound metrological infrastructure (metrology is the science of fine measurement) so that governments and manufacturing, commerce, health and safety, and other sectors can have access to a wide range of measurement, calibration and testing services in which they can have complete confidence. Figure 6.16 shows the structure of the national measurement system that is in place in the UK. It shows that the NPL is at the focus of the system and maintains the national primary standards both for the seven SI base units (such as mass, length and time) and the SI supplementary units (such as force and electric potential difference).

Commerce, industry and other users do not have direct access to the primary standards, but they have access to secondary standards that have, themselves, been calibrated against the primary standards. NAMAS provides an essential part of this hierarchical structure through accredited laboratories and test houses, which give their customers access to authenticated measurements, calibrations and tests of all kinds. These have formal and certificated traceability to the national primary standards at the NPL. Confidence in the accuracy of measurement and test data is a vital part of ensuring product quality.

Manufacturing industry is constantly being encouraged to improve, in every aspect of its quality procedures, and NAMAS plays its part by emphasising the importance of trustworthy measurements and tests.

On an international scale, laboratories, test centres and inspection departments accredited by a body that is trustworthy and whose competence is beyond question, makes the products and services of a country more readily acceptable overseas. This is particularly important with the advent of the Single European Market. For this reason, NAMAS cooperates fully with other international agencies concerned with laboratory accreditation. Mutual agreements recognising each other's competence are negotiated where appropriate.

6.23 Total quality management

Quality is not only about 'good design' and 'conformance to design specifications'. Quality should permeate every aspect of a company's business, always focusing upon the needs and hence the care of the customer. The traditional approach to quality has been that 'production makes it' and 'quality control inspects it'. This approach leads to financial loss being incurred every time defective products are scrapped or time is involved in reworking and reclaiming defective products. This also leads to late delivery and dissatisfied customers who have to be pacified, and who have the same problems with their own customers. It has been estimated that as much as two-thirds of all the effort utilised in business is wasted by this domino effect. Thus the objectives of any company can be defined as:

- optimum quality;
- minimum cost;
- shortest delivery time accurately adhered to.

To achieve and, more importantly, maintain these objectives will require something better than the traditional approach to quality. It will require more than an attempt to 'inspect-in' quality. It requires the efforts of every resource: the total commitment of every member of the company to satisfying the customers' requirements at all times. The slogan 'quality is everyone's business' just remains a slogan if it is not backed by positive action throughout the whole of a company's activities. It can soon change to 'quality is nobody's business' without managerial determination from the board of directors downwards. From this approach of total involvement has developed the concept of *total quality management* (TQM). Total Quality Management can be defined as:

> an effective system of co-ordinating the development of quality, its maintenance and continuous improvement, with the overall objective of the achievement of customer satisfaction, at the most economical cost to the producer and, hence, at the most competitive price.

In theory, since quality affects everyone in the factory, it can be truly stated that 'Quality is everyone's business'.

6.24 Quality assurance: BS EN ISO 9000

Although, under TQM, everyone from the boardroom down is responsible for the achievement of quality standards, the 'buck stops' at the desk of the company officer who signs the quality assurance certificate. Quality assurance (QA) is defined in BS 4778: 1987 as 'all activities and functions concerned with the attainment of quality'. In practice, it is some formal system in which all the operations are set out in written form as procedures and work instructions. The documents are subject to regular review and update and are usually under the authority of one person, the *quality manager*.

With the introduction of TQM, products and services not usually associated with certification have been brought into line by there being written evidence that the company has achieved nationally accredited standards of quality. These standards are accepted internationally because the standards of accreditation are common among all the industrial communities. International interest in quality standards led to the introduction of ISO 9000: 1987. This has been adopted without alteration by the British Standards Institution as BS EN ISO 9000. The new ISO standard also incorporates the European Standard EN29000.

6.24.1 *The basis for BS EN ISO 9000*

The definition of quality upon which this standard is based is in the sense of *fitness for purpose* and *safe in use*, and that the product or service has been designed to meet the customer's needs. The constituent parts of this standard set out the requirements of a *quality-based* system. No special requirements are incorporated that might be only needed or even achievable by a few companies. BS EN ISO 9000 is essentially a practical basis for a quality system or systems that can be utilised by any company offering products or services both within the UK or abroad. The principles of the standard are intended to be applicable to any company of any size. The basic disciplines are identified with the procedures and criteria specified in order that the product or service meets with the requirements of the customer. The benefits of obtaining BS EN ISO 9001/2 approval are very real, and they include:

- cost effectiveness because a company's procedures become more soundly based and criteria more clearly specified;
- reduction of waste and the necessity for reworking to meet the design specifications;
- customer satisfaction because quality has been built-in and monitored at every stage before delivery;
- a complete record of production available at every stage to assist in product or process improvement.

By the adoption of BS EN ISO 9001/2, a company can demonstrate its level of commitment to quality and also the ability of that company to supply goods or services to the defined quality needs of customers. For many companies it is merely the formalising and setting down of an existing and effective system in documented form so that the validity of the system can be guaranteed by obtaining *external accreditation*.

By having an agreed standard as the basis of supply there can be major benefits to all the parties concerned. A customer can specify detailed and precise requirements, knowing that the supplier's conformance can be accepted. This is because the quality system has been scrutinised by a third party, i.e. it has been *accredited*. However, successful implementation will be dependent upon the total commitment of the whole management team and, in particular, the person at the very top of the organisation.

6.24.2 *Using BS EN ISO 9001/2*

Customers may specify that the quality of goods and services they are purchasing shall be under the control of a management system complying with BS EN ISO 9001/2. Together with independent third parties, customers may use the standard as an assessment of a supplier's quality management system and, therefore, the ability of a supplier to produce goods and services of a satisfactory quality. The standard is already used in this way by many major public-sector purchasing organisations and accredited third party certification bodies. Thus it is in the interests of suppliers to adopt BS EN ISO 9001/2 in setting up their own quality systems.

The direct benefits to companies who have been assessed in relation to the standard and who appear in the DTI's *Register of Quality Assessed United Kingdom Companies* consist of reduced inspection costs, improved quality, and better use of manpower and equipment.

Where a company is engaged in exporting goods or services, there is a direct advantage in possessing mutually recognised certificates that could be required by overseas regulatory bodies.

6.25 Control of design quality

The objective of any design is to meet a customer's requirements in every respect. Therefore, it is vital that the input to the design function is established and documented. This input is the result of consultation by the marketing and the design staff with the customer. This can easily be achieved by use of the *Internet* and *3D modelling* (see Chapter 1) if the purchaser and the supplier are geographically separated. There should be sufficient trained staff and resources available to ensure that the design output meets a customer's requirements, and that any changes or modifications have been controlled and documented on computer files.

6.25.1 *Traceability*

All the products needed to fulfil a customer's requirements should be clearly identified in order that they can be traced throughout the company. This is necessary in order that a capability exists for the tracing of any part that could be delivered to a customer. The need for this traceability might arise in the case of a dispute regarding non-conformity or for safety or even statutory reasons. Identification is also important where slight differences exist between the requirements of different customers.

6.25.2 *Control of 'bought-in' parts*

All purchased products or services should be subjected to verification of conformance to previously agreed specifications. Remember that, usually, all organisations are both customers and suppliers. Control is exercised by documentation of the purchasing requirements, together with the inspection (and hence verification) of the purchased product. Many companies insist on receiving details of the quality system used by subcontractors and suppliers. Hence companies can obtain value by being accredited under BS EN ISO 9001/2. Even where a customer supplies a product to be processed, the customer must be assured that the receiving company is, itself, 'fit for purpose' to carry out the process. While a product is in the possession of a processing company, that company is responsible for its well-being, i.e. that it remains free from defects.

6.25.3 *Control of manufacturing*

Clear work instructions are at the heart of any manufacturing process. They eliminate any confusion by showing, in a simple manner, the work to be done or the service to be provided and also indicate where the responsibility and authority lie. If a customer's requirements, via the design function, are clearly specified on the work documentation together with quality criteria, then the task of the manufacturing function becomes more easily controlled. British standard BS EN 9001/2 spells out the items that work instructions should include, and emphasises the control of any additional processes, such as heat-treatment, thus avoiding expensive errors.

6.25.4 *Control of quality of manufacture*

Prompt and effective action is essential to the maintenance of quality throughout the system. Not only must defective parts be identified but the causes must quickly be rooted out. This could lead to changes in the design specifications or the working methods. Control of the manufacturing process starts with inspection of the incoming goods for conformance to agreed specifications. Documentary evidence of such conformance should be provided either by the external supplier or by the company's own in-coming inspection. The accreditation of every company's measuring and testing equipment by an external body such as NAMAS provides consistency and avoids disputes between supplier and user.

An essential part of any quality control system is the means of indicating the status of goods with regard to their inspection, that is 'inspected and approved', 'inspected and rejected, or 'not inspected'. Any part that does not conform to the specifications must be clearly identified to prevent unauthorised use or dispatch. Documentation must be raised to indicate the nature of the non-conformation and of either its disposal or any remedial work performed upon it.

6.25.5 *Summary*

British standard BS EN ISO 9001/2 is about documentation procedures and, as previously stated, *is ideally suited to computerised systems*. It is not a panacea for quality but it is a mechanism by which a company can formalise its operation in a standardised way. The computerised procedures must be unambiguous and clearly state what is to happen, **not** what the company thinks ought to be happening! The system chosen should be compatible with the business and should be capable of being maintained. Communication throughout the organisation is essential. The workforce must know what is happening and how it will affect them. This is particularly important during the assessment period, when accreditation is being applied for, as the assessors may talk to anyone in the company. It is important, subsequently, to maintain and improve upon the standards achieved during the accreditation assessment.

6.26 Methods of inspection

There are basically two methods of inspection that can be carried out at any stage of a process. They are:

6.26.1 *Total inspection*

With 100 per cent inspection every part will be examined. It may be thought that this will guarantee that all defectives will be identified. However, this is not necessarily the case where human beings are the inspectors and arbiters of conformance. This type of monotonous, repetitive inspection procedure leads to loss of concentration and the missing of some defectives. This can be as high as 15 per cent at the end of a shift. Some products themselves preclude 100 per cent inspection. For example the ultimate test of a ballistic missile lies in the actual firing! Therefore, some other means must be found.

6.26.2 *Sampling*

With sampling, a decision on quality is based upon the close inspection of a randomly selected batch of materials or products. The decision as to whether the remaining materials or products are acceptable or not is based upon the outcome of mathematically-based statistical procedures. The process involves the collection of data and making decisions about conformance as a result of plotting the collected information on various control charts.

Whichever system of inspection is adopted, it should be clear that to employ only 'end-inspection' is wasteful and uneconomical. In many cases it is far too late, resulting in delayed deliveries and either total rejection of a batch of work or, at best, considerable reworking to reclaim the batch at considerable extra cost. Inspection should be used as a check on a process while it is in progress and add value to a job. It is most effective when it is in the hands of the production operatives, with the quality control section only acting in a validating role. This is the principle that *prevention is better than cure.*

6.26.3 *Position of inspection relative to production*

Consideration must be given to the position of any in-process inspection points in order that costly time is not wasted on processing parts that are already defective. Such inspection points could be:

- prior to a costly operation (e.g. there is no point in gold-plating a defective watch case);
- prior to a component entering into a series of operations where it would be difficult to inspect between stages (e.g. within a flexible manufacturing system cell);
- prior to a station that could be subjected to costly damage or enforced shut-down by the failure of defective parts being fed into it (e.g. automatic bottling plants, where the breakage of a bottle could result in machine damage or failure, in addition to the cost of cleaning the workstation);

- prior to an operation in which defects would effectively be masked (e.g. painting over a defective weld);
- at a point of no return where, following the operation, rectification becomes impossible (e.g. final assembly of a 'sealed for life' device).

6.27 Areas of quality control

In all manufacturing organisations there are three main areas of quality control. Let us now consider them individually.

6.27.1 *Input control*

It would be foolish and wasteful to utilise resources on the processing of materials or parts that are already defective when received from the supplier. The function of incoming inspection is to detect any faults that could affect the finished product. An example of this could be the machining of rotor and turbine shafts for the electricity generating industry. The very large forgings for the shafts are inspected internally and externally for cracks before being subjected to extensive and costly machining processes.

One way of effecting input control is to rely upon the supplier's out-going inspection. This could mean the production of documentary evidence in the form of certificates, hence the value of BS EN ISO 9001/2 in standardising quality documentation.

Alternatively, the customer has two other courses of action. One course of action is for the customer to provide supervised inspection at the supplier's plant prior to despatch. The other is for the customer to carry out 'in-company inspection' at their own plant on receipt of the goods or materials, but before accepting a delivery. This could be 100 per cent inspection or sampling to arrive at a decision regarding acceptance or rejection.

6.27.2 *Output control*

This is the last link in the quality control chain. It is necessary not because the previous levels of inspection have failed, but because of the fallibility of any system of inspection and the resulting consequences of the customer receiving products or services that do not conform to the agreed specification. Again the choice is between 100 per cent inspection, or sampling.

6.27.3 *Process control*

This is the 'let us see how it is doing' approach and, if it is successful, reduces the need for output control to a minimum. Considering that any process has an inherent amount of variability, it is important that the amount of this variation is known. Thus a capability study has to be performed and the process has to be kept within acceptable limits. The problem for *process quality control* is, therefore, how to detect unacceptable process variation before it leads to non-conformance with the agreed specification.

6.28 Quality control techniques

Quality assurance can be defined as, 'the provision of documentary evidence that an agreed specification has been met and the contractural obligations have been fulfilled'.

It is a powerful definition indicating that all the legal implications associated with the law of contracts have been covered. Quality control (QC) can be defined as, 'the application of a system for programming and coordinating the efforts of the various groups in an organisation to maintain or improve quality at an economical level that still allows for customer satisfaction'.

The application of these techniques and procedures is necessary to ensure that any goods or services actually meet the design specifications and hence the customer's contractual requirements.

Thus, QA is the objective to be attained, and QC is the means of achieving that objective. Before any type of control can be exerted in any sphere, ground rules have to be established. In the case of QC, in common with any type of control, those ground rules consist of:

- the setting of standards against which performance can be compared;
- the mechanics for assessing performance against the prescribed standards;
- the means of taking any necessary corrective action where an error exists between actual performance and the prescribed standards, as shown in the quality control loop, Fig. 6.17.

Fig. 6.17 *The quality control (QC) loop*

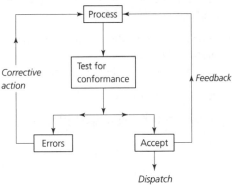

6.29 Quality control in e-manufacture

For a full and detailed account of *quality control in manufacture* together with statistical sampling, the reader is referred to *Manufacturing Technology*, Volume 2, by the authors of this book. However, this brief account should show how information and communications technology is an ideal vehicle for:

- processing the documentation involved;
- processing the sampled data and performing the calculations involved;

- providing outcomes for process control;
- liaison with customers and suppliers from the design stage to final delivery in order to achieve products of the required quality (*fitness for purpose*).

In addition, e-manufacture is an ideal means of achieving the required quality standards and maintaining customer satisfaction, in that it:

- removes or minimises labour costs;
- removes the chance of human error;
- improves consistency of quality;
- reduces the need for costly inspection processes if adaptive control and linked, automated, coordinate measuring machines are used.

ASSIGNMENTS

1. Consider a company, business or institution you work at or attend and list the documentation that could become available on an intranet.

2. List the advantages and disadvantages of this new system over a paper-based system.

3. Visit a few commercial Web sites and consider the advantages in terms of marketing and other business functions. Give examples of material contained on these sites that enhances business function.

4. Give examples of on-line communities on the Web; explain the benefits of such communities.

5. Explain how computers have increased the efficiency of manufacturing industry. Also analyse the downside, i.e. are we becoming slaves to the machine and in some instances using a sledgehammer to crack a nut?

6. Explain how quality may be enhanced using the latest computerised systems.

7. Analyse the attributes of a good Web page interface.

7 The future of manufacturing

When you have read this chapter you should be able to:

- make considered judgements as to the future of manufacturing technology;
- question existing methods of manufacture;
- look for improved methods of manufacture;
- assess the value of new processes;
- assess the depreciation costs and profitability of new plant;
- analyse new concepts such as parallel kinematic mechanisms;
- appreciate the advantages of intelligent machines;
- appreciate the advantages of miniaturisation;
- appreciate the need for and methods of security in e-manufacturing and e-commerce;
- appreciate the techniques for part-sourcing in a global market;
- appreciate the importance of green process planning (GPP) technology.

7.1 Introduction

Ever since the beginnings of the 'industrial revolution', engineers have been searching for ways of automating manufacturing processes. They have done this for a number of reasons, such as to:

- reduce manufacturing costs by reducing the dependency of manufacturing processes on human labour;
- increase the volume of output per unit time to satisfy consumer demand;
- ensure that manufactured goods are available as and when required;
- improve the repeatability and accuracy necessary when manufacturing interchangeable, standard components;
- prevent contamination by contact with human beings;
- enable dangerous substances to be manufactured without exposing human beings to hazardous environments;
- prevent environmental disasters occurring through human error.

Up to the middle of the twentieth century, automated machines were dependent on interchangeable cams and similar devices for their control. Towards the end of the Second

World War, the programmable computer was developed. This was based on thermionic valve technology and as a result it was cumbersome, limited in its memory capacity and, by later standards, slow and unreliable. It was the 'sea-change' in technology that followed the invention of the transistor and, soon afterwards, integrated circuits that resulted in compact but powerful computers. Without this development there would be no e-commerce or e-manufacturing.

To some extent this is a 'chicken and egg' situation. Without computerised manufacturing processes and without pick and place robots it would be impossible to make the miniaturised components and assemble them in position in a computer to enable the speed and accuracy that are required. Yet without computers and computerised technology, computerised manufacturing and management would not exist.

The development of transistor technology from which computer aided manufacture, and computer aided management and commerce, stems could not have been predicted. It is not possible to know what the next 'sea-change' in technological development will be and what it will hold for the future of manufacturing. Therefore this chapter is not an attempt to predict what the future holds for the technology of manufacture and its management. It is an attempt to look at some trends and how they are influencing change in manufacturing processes and management in many sectors of industry.

7.2 Investment in new plant – making considered judgements

In the earlier chapters of this book we have considered a number of changes and developments that have become possible through the adoption of computer technology. Although some technologists and hardware engineers are of the opinion that the speed and memory capacity of computers is thought to be reaching the limits of development, software engineers are constantly developing new applications around the existing technology. Therefore changes and development of information and communications technology (ICT) applications to manufacturing processes and management will not cease. However, frequent change brings its own problems. The time has long since passed when capital plant such as machine tools could be purchased with a life expectancy of half a century or more. In fact the downfall of manufacturing in the UK has been largely due to resistance to change by the workforce for sociological reasons and by management for largely economic reasons, but also through lack of understanding of the benefits that ICT can bring. Unfortunately automated plant is very much more costly than manually controlled plant. There is a saying that the PC you have just bought is out of date by the time you have carried it out of the shop. The same applies to ICT controlled manufacturing plant and systems. Therefore it is necessary to make carefully considered judgements on the future of manufacturing technology before investment is made in new manufacturing plant. It is necessary to give careful thought not only to the cost involved but also to the working life of the equipment before it becomes obsolete through advances in technology, truly a case of *caveat emptor*.

7.2.1 *Depreciation*

Depreciation is the reduction in the intrinsic value of an item of plant or equipment due to normal wear and tear or due to the fact that it has become technologically out of date. It must be remembered that each item of plant represents a capital asset in a company's accounts. Since the value of each item of plant is continually diminishing, this represents a reduction of the capital assets of the company and must be shown in the balance sheet if this is truly to represent the financial status of a company. Furthermore, a provision must be made for the eventual replacement of the items of plant under consideration. To avoid repeatedly using the phrase 'item of plant' throughout this section, the term 'asset' will be used instead. The original cost of each asset *plus* all its additional expenditure (such as servicing and repairs and maintenance) *less* its residual 'trade-in' or scrap value, must be charged against revenue (the money earned by that asset). This charge must be spread over the asset's economic service life as fairly as possible. Remember that an asset's service life is usually very much shorter than its actual useable life. This is particularly the case where computer hardware and software are involved in process control. Furthermore, the trade-in value of computer hardware and software is negligible, so the cost of computers and computerised machines and equipment must be written off over a very short period of time, certainly not more than five years. To provide for the replacement of an asset, a sum of money must be set aside annually from revenue earned by the asset so as to accumulate the required amount when the end of its service life is reached.

Allowance must also be made for inflation. It is essential to create this *depreciation reserve fund* so that new capital does not have to be raised when replacement becomes necessary. From time to time technological innovation may make it necessary to replace plant prior to the planned date and before adequate funds are available. Under such circumstances it may well be necessary to raise additional capital and the cost of servicing such capital (loan costs) must be included in the estimates of revenue the new plant will earn over its lifetime. There are various methods of writing down the value of an asset over its lifetime and some of these will now be examined. For simplicity the fact that the depreciation funds can be invested to accrue interest (compound) will be ignored but it can be helpful to some extent in offsetting the effects of inflation.

7.2.2 *Fixed instalment method*

This is also known as the 'straight line' method since the value of plant is written down in equal amounts over equal increments of time, thus producing a straight-line graph. Let I equal the initial cost, R equal the remnant (final disposal) value, and n equal the number of years over which the value of the item of plant is being written down. Then, the reduction in asset value and the sum set aside each year should be $(I - R)/n$. This is shown in Fig. 7.1. This simple method makes no allowance for the fact that the cost of repairs and maintenance is greatest at the end of the life of an asset.

7.2.3 *Reducing balance method*

This allows for the fact that, financially, most assets depreciate most rapidly in the early years of their life and that the depreciation – as a percentage of the original purchase

Fig. 7.1 *Fixed instalment (straight-line) depreciation*

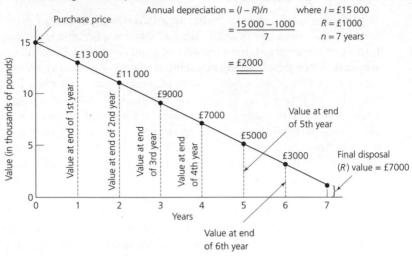

Annual depreciation = $(I - R)/n$ where $I = £15\,000$
$$= \frac{15\,000 - 1000}{7} \qquad R = £1000$$
$$n = 7 \text{ years}$$
$$= £2000$$

Fig. 7.2 *Reducing balance (%) depreciation*

Year	Start of year value (£)	Depreciation (£) ($n = 32\%$)	End of year value (£)
1	15 000	32% of 15 000 = 4 800	15 000 – 4 800 = 10 200
2	10 200	32% of 10 200 = 3 264	10 200 – 3 264 = 6 936
3	6 936	32% of 6 936 = 2 221	6 936 – 2 221 = 4 715
4	4 715	32% of 4 715 = 1 509	4 715 – 1 509 = 3 206
5	3 206	32% of 3 206 = 1 026	3 206 – 1 026 = 2 180
6	2 180	32% of 2 180 = 698	2 180 – 698 = 1 482
7	1 482	32% of 1 582 = 474	1 482 – 474 = 1 008

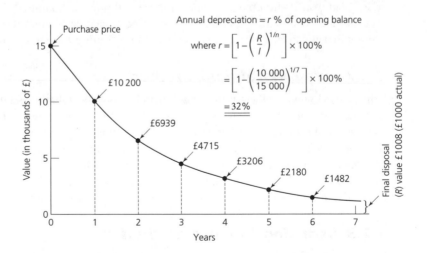

Annual depreciation = r % of opening balance

where $r = \left[1 - \left(\dfrac{R}{I} \right)^{1/n} \right] \times 100\%$

$= \left[1 - \left(\dfrac{10\,000}{15\,000} \right)^{1/7} \right] \times 100\%$

$= 32\%$

price – becomes less each year. Using the symbols from the previous example, and letting r equal the fixed percentage, then r can be determined from the expression $r = [1 - (R/I)^{1/n}] - 100$. The application is shown in Fig. 7.2.

7.2.4 Interest law method

This provides for depreciation by crediting the asset account with equal yearly amounts. The asset account is then debited with the interest that would otherwise be earned by the capital invested in the asset had that capital been invested in a bank deposit account. The notional interest so deducted is then credited to the *profit and loss account*. The annual increments, less the notional interest on the yearly *balance*, indicates the amounts to be set aside for replacing the asset at the end of its forecasted service life.

Let S equal the sum set aside each year. Then, using the previous symbols,

$$S = [P(I - R)]/[(1 + P)^n - 1]$$

where P is the notional rate of deposit account interest. This method of calculating the depreciation allowance is frequently used in connection with asset leasing. Whichever of the three previous methods is used, the accumulated reserve at the end of the service life of the asset (nth year) will be $I - R$, as shown in Fig. 7.3.

7.2.5 *Miscellaneous methods of assessing depreciation allowance*

There are other systems of allowing for depreciation. For instance, in the 'depreciation fund' system the asset is shown at its full value throughout its life but the *profit and loss account* is debited with a fixed annual sum that is paid into a deposit account or into gilt-edged securities. The 'insurance policy' system is similar to the previous example except that the annual sum debited from the profit and loss account is credited to an insurance policy that secures a sum sufficient to replace the asset at the end of its service life. The 'revaluation' system requires each asset to be revalued each year. A charge equal to the difference between the book value and the assessed value is then debited against revenue earned by the asset and placed in a sinking fund ready for eventual replacement of the asset, the book value being duly written down. An independent professional valuer is required in the interest of accuracy and impartiality. Finally, in the 'single charge' system, a single charge is made annually against revenue earned by the asset to cover repairs, renewals and depreciation. The charge may vary and revaluation may also be required from time to time.

7.2.6 *Break-even analysis*

This is another useful tool when making a considered judgement on purchasing a new item of plant: the most important consideration is whether or not the plant will make an adequate profit. Therefore, in addition to considering the depreciation of any new plant, it is necessary to carry out a break-even analysis. An example of a break-even graph is shown in Fig. 7.4. This is a simple graph that, for the moment, assumes the product can be made on the existing plant. The vertical axis represents *cash* both as

Fig. 7.3 *Interest law method of calculating depreciation*

Year	Sum set aside each year (£)	Balance carried forward (£)	Accrued interest (£) at 8%		Balance at year end (£)
	(S)				
1	1 569	–	–		1 569
2	1 569	1 569	8% of	1 569 = 126	3 264
3	1 569	3 264	8% of	3 264 = 261	5 094
4	1 569	5 094	8% of	5 094 = 408	7 071
5	1 569	7 071	8% of	7 071 = 655	9 206
6	1 569	9 206	8% of	9 206 = 737	11 512
7	1 569	11 512	8% of	11 512 = 921	14 007

$$S = [p(I - p)] / [1 + p)^7 - 1] \qquad \text{where: } p = 8\% \text{ interest}$$
$$= [0.08(15\,000 - 1000)] / [(1 + 0.08)^7 - 1]$$
$$= £1569$$

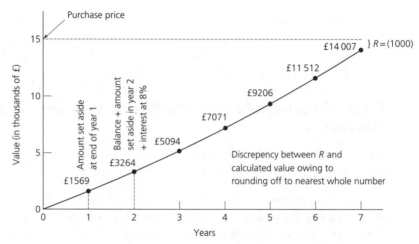

revenue from sales and as the cost of manufacture. The horizontal axis represents units of production (number of parts made). Line AB represents the money earned by the sale of the parts made. The greater the quantity of parts made and sold the greater the revenue earned. The line AB is drawn from the origin of the axes because if no parts are sold no revenue will be earned. The cost of manufacture is rather more complex and needs to be built up from the fixed costs (overheads) and production costs (variable costs). The overheads can be considered to be virtually constant irrespective of the number of parts made, and represent such items as rent, rates, office salaries, plant and telephones, together with electricity and gas for lighting and space heating, as represented by the line CD. The cost of production (variable costs) represents such items as materials, direct labour (wages), electricity for powering production equipment and gas for process heating. This has to be added onto the overhead costs and is represented by the line CE. The variable production costs are proportional to the number of parts made. Double the number of parts and you double the variable production costs. Where the lines AB and CE cross is the *break-even point*. At this point the income and expenditure are equal. If the number of parts made and sold are below this quantity then they will be made at

Fig. 7.4 *Break-even graph (1)*

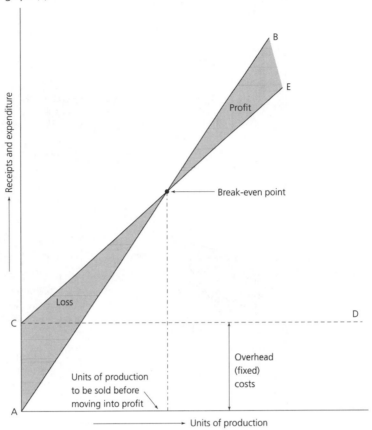

a *loss*. If the number of parts made and sold are above this quantity then they will be made at a *profit*.

To increase production and to improve product quality, management may decide to invest in a CNC machine to make the parts concerned. Because of the cost of the investment in the new item of plant and the need to cover its depreciation, as previously discussed, there is a corresponding, substantial increase in the overhead item PQ in Fig. 7.5. At the same time, the variable cost of production will be reduced by eliminating the direct labour cost element. However, there will still be material costs in proportion to the number of items made and there will be increased maintenance and skilled setting costs, so we cannot eliminate all the variable cost item, only reduce it, as shown by the line PR. If the selling price is maintained then the break-even point will be reduced and the process will become more profitable. Alternatively the selling price, as shown by the line AE in Fig. 7.6, can be reduced to make the product more competitive. However, the overhead and investment costs are unchanged so a great many more units of production must be sold before the maker moves into profit. Invariably a compromise has to be reached between profitability and competitiveness. Furthermore, it is no good increasing the number of products that need to be sold before moving into profit if the market cannot absorb them.

Fig. 7.5 *Break-even graph (2)*

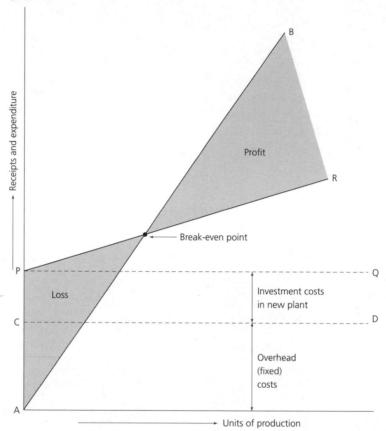

The graphs only represent the situation at a particular instant in time. Any active company will be constantly adjusting the slopes of the various elements of the graphs to maximise its profits and maintain or improve its competitiveness. Any increase in overheads resulting from the purchase of new and, hopefully, more productive plant should be more than offset by a reduction in direct labour costs. The company should also be looking constantly for cheaper and better materials. The selling price will constantly need to be tuned to the maximum the market will stand. If new plant has to be bought, then the cost of servicing loan charges and the depreciation allowance have to be added into the equation. The reality is, therefore, much more complex than the previous, simple example. Complex mathematical modelling using sophisticated accountancy software is required, yet another example of e-commerce supporting e-manufacture and another task for the computer.

7.2.7 New plant – the decision

By now it should be clear that the purchase of a new item of plant cannot be taken lightly. To have surplus capacity 'just in case it is needed' or because it represents 'cutting edge

Fig. 7.6 *Break-even graph (3)*

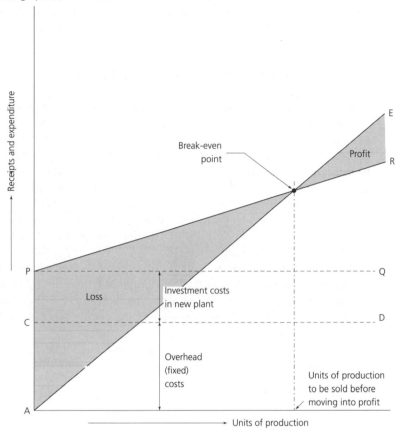

technology', is a luxury that can no longer be afforded when companies have to be lean and efficient to survive in highly competitive global markets. The only time a costly new item of plant can be justified is when it can be worked to its full capacity and profitability, to pay off the purchase cost before the plant becomes obsolete. This will invariably mean keeping the machine fully loaded throughout its useful life. If this cannot be achieved then it is usually better to sub-contract the product to a specialist manufacturer who has the necessary specialist plant and a sufficient volume and variety of work to keep that plant continuously loaded.

7.3 The future

There is no excuse for not challenging the use of existing production equipment and methods of manufacturing. In fact, a complacent mind-set has resulted in the economic demise of many established companies. For example, many manufacturers of steam locomotives went out of business through being incapable of moving into the era of diesel and electric traction. It is as important for manufacturing engineers to be constantly on the lookout for improved methods of manufacturing, as it is for design engineers to keep the firm's products up to date. This is the only way that their firms can remain in business. Let us now consider some examples that may lead us towards the future of manufacture.

7.4 Parallel kinematic mechanisms

Earlier in this book we considered the application of programmable electronic control to a range of processes. Although the control devices themselves represent state of the art technology, to a large extent the devices being controlled represent technology that has been tried and tested over many years, and they have not changed significantly from when they were manually controlled. In other words we have merely exchanged the human operator for a CNC or a PLC or a PC control unit. The basic machines have undergone constant development in an attempt to remain competitive but the underlying principles of design and construction have changed very little. We are now going to look at a concept machine developed originally by Giddings and Lewis of America in conjunction with Nottingham University in the UK.

Conventional machines are built up from heavy, robust castings supported on large slideways in order to withstand the loads imposed by the cutting forces, the mass of the workholding devices and the mass of the workpiece itself. Furthermore, this form of structure damps out the vibration caused by the cutting process and ensures repeatable accuracy, but is it the best solution?

The *Variax concept milling* machine developed by Giddings and Lewis of America takes a fresh look at machine tool design and construction. It incorporates one of the most notable developments in machine tools as a direct result of the introduction and application of information technology to machine tool design, namely, the adoption of *parallel kinematic mechanisms* (PKMs). Many parallel structures are based upon the '*Stewart Platform*', a term synonymous with any machine based upon parallel kinematics. Parallel kinematic mechanisms have been known for some time and were applied to tyre testing machines and flight simulators as far back as the 1960s. However, it required the availability of computing power at the beginning of the new millennium to make the application of PKMs to machine tools a reality.

It was not until control systems became powerful enough to cope with solving the required transformation equations for PKMs in real time that an interest in parallel kinematics was rekindled. The 'Delta' robot (with three degrees of freedom) was the first commercially available system, and was later developed into the 'Hexa' with six degrees of freedom in 1991. These systems have high speeds and accelerations and are primarily used in the packing industry for pick and place applications. In 1994 the first PKM machine tools began to emerge when Giddings and Lewis launched the *Variax Hexacenter*. Other machine tool manufacturers have experimented with hybrid technology combining some elements of PKMs with elements of conventional machines.

7.4.1 *PKM definitions*

A conventional machine tool will consist of separate drives for each axis of motion. Slideways are used to ensure accurate movement along the linear X-, Y- and Z-axes

Fig. 7.7 *The Stewart Platform (simplified representation of a PKM)*

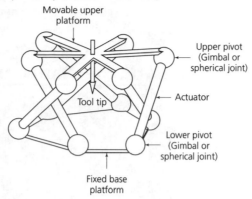

Movable upper
platform

Upper pivot
(Gimbal or
spherical joint)

Actuator

Tool tip

Lower pivot
(Gimbal or
spherical joint)

Fixed base
platform

and a combination of a rotating head or table can be used as a pivot for additional rotational axes A (rotation around X), B (rotation around Y) and C (rotation around Z). A convenient term for a combination of drive unit and slideway, or drive unit and pivot, is a 'device'. When devices are stacked one on top of the other, as in a conventional machine tool, this arrangement is called '*serial kinematics*'. In this context conventional machine tools, both manual and CNC, are most accurately described as '*serial kinematic mechanisms*' (SKMs). Each element of a conventional machine has six degrees of freedom that must be constrained by slideways and controlled by traverse screws and nuts. If a machine is built up from, say, three stacked elements in series then there are $(3 \times 6) = 18$ sources of error.

The alternative method of producing motion is to use '*parallel kinematics*', where the devices are not arranged on top of each other and are not aligned to conventional axes. Such parallel kinematic devices, lacking the constraints of conventional machine tool slideways, must be precisely controlled using computer technology to implement the complex mathematical transformations involved. Figure 7.7 shows a simplified representation of one of the many possible configurations of a 'parallel kinematic mechanism'. As stated earlier, in addition to PKMs and SKMs there are also 'hybrid' machines that have a mixture of parallel and serial elements.

The conventional method of indicating the accuracy of a machine tool was to evaluate the linear accuracy of each axis. As accuracy requirements increased, the alignments of these axes became an important issue and the tolerance of these axes was also quoted. As profiling became a common operation for machine tools, a new type of measuring accuracy was used: 'volumetric' accuracy. This is commonly used in co-ordinate measuring machines (CMMs), which are often required to work volumetrically as the component to be inspected is not required to be aligned to the axes.

7.4.2 *Principle of operation*

Conventional machine tools control the motions of the workpiece and/or the cutter by means of substantial and accurately aligned slideways, actuated by leadscrews and nuts. The accuracy of the work produced by such machines is very largely dependent upon

the geometry and stiffness (freedom from deflection) of the relevant structures. Most users are familiar with the effect of backlash, stick-slip and deflections under cutting loads. Limitations also include such parameters as spindle speeds and available cutting power. No matter how ingenious the configuration, conventional machine tool design is normally constrained by these features. Accuracy depends upon rigidity and rigidity depends upon mass.

Conventional machines that operate under computer control are normally equipped with servomotors to rotate the necessary leadscrews. Servomotors can be very powerful and responsive to the control of acceleration and deceleration. The leadscrews themselves are invariably of the recirculating ball type. This reduces wear, backlash and friction and ensures minimum clearances and maximum accuracy. A degree of pre-loading is also possible.

The Variax Hexacenter replaces the massive perpendicular linear slideways and bearings of conventional machine tools with a system of six crossed legs connecting a lower platform – carrying the workpiece – to an upper platform carrying the spindle and cutting tool, as shown in Fig. 7.8. The six legs that control the motion of the tool contain ballscrews and provide an immensely strong structure. The real secret of the system is the mathematical analysis of the relationships between the six actuators that control the ball screws and move the tool to the required position. So far as the user is concerned the machine responds to commands related to the conventional six degrees of freedom, namely linear motion along the three mutually perpendicular X-, Y- and Z-axes and rotations (A, B and C) around each axis. The Variax machine control system converts commands presented in such terms into extension or contraction of the six legs thereby providing the relative tool/workpiece movements. Only the development of powerful computers and software has made this possible. During machining the six legs are constantly extending or contracting to guide the tool over the workpiece and provide the contour required. The movements of the legs are constantly monitored by means of laser measuring units (encoders).

It may be thought that such a machine would lack the rigidity of the conventional machine. However, the Variax Hexacenter is a 'smart' machine. As well as controlling the struts that provide the desired cutter path, the encoders also detect any deflections of the struts caused by the cutting forces. Corrections are computed and the strut servos make the necessary compensating adjustments. This happens in milliseconds so that the machine is constantly self-aligning.

7.4.3 *Advantages of PKMs*

No alignment required

It is not a requirement of PKMs that the actuators have to be aligned accurately because the calibration routine can identify their location. This is a benefit because high cost, accurately finished castings, labour-intensive alignment procedures and accurate and expensive workholding fixtures are not required.

High volumetric accuracy

A well-designed PKM can achieve very high volumetric accuracy because the errors from each actuator do not build up as in a serial structure; instead, an averaging effect is seen at the tool tip. The key to this statement is 'well-designed' since is easy to create a PKM with very low accuracy. A closed-loop feedback control system using glass scales, laser interferometers or similar devices to monitor the motion of the actuators is required. Thermal effects and structural deflections must be monitored and minimised, and the rotational accuracy of pivot points (sphericity) of a joint or the alignment of a gimbal's axes is also critical.

High rigidity

A well-designed PKM can also achieve very high static and dynamic rigidity as forces are transferred through the structure creating only tension and compression in the actuators. This means that bending moments, which impair both accuracy and rigidity, can be eliminated.

High velocity

Most motions of a PKM require all the actuators to be operating at the same time. This means that the velocity at the tool tip is usually higher than the velocity of any of the individual actuators. It is important to note that although this holds true for three-axis contouring, it is not necessarily the case with five-action motion. Another advantage is that the cutter axis is always normal to the surface being cut, producing a relatively smooth contour. In a conventional machine, where the contour is produced with the cutter always perpendicular to the worktable and the contour is generated in a series of steps, hand finishing is required to smooth them out. This is shown in Fig. 7.9.

High thrust

A PKM can exert a high thrust for the same reason that high velocity can be achieved: all six actuators work together to produce the motion.

Fig. 7.9 *Comparison of contouring when using (a) a conventional CNC milling machine and (b) the Variax Hexacenter*

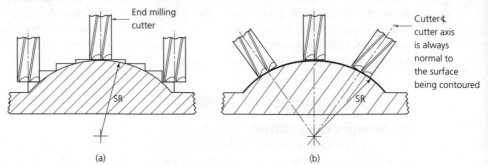

Low mass

Since bending moments are eliminated it is possible to design a machine with a lower mass relative to a similar, conventional machine, which will often have large, heavy castings. This allows high rates of acceleration and deceleration to be achieved in the movement of the machine elements. It also avoids the need for heavy lifting equipment during manufacture, heavy transport between the foundry and the machine tool manufacturer, and reduced amounts of material extraction, which, in turn, reduces environmental pollution.

High acceleration and jerk

The combination of high thrust and velocity with low mass leads to the potential for high acceleration and 'jerk' (rate of acceleration).

Reduced foundation requirements

Lower mass also reduces the foundation requirements. However, if the machine has higher acceleration than a conventional machine, this may not always be the case because the increased momentum would result in greater energy transfer to the foundation.

Cost

This is a contentious issue because most PKMs are still in the early stages of development; as such they tend not to be mass-produced and have a correspondingly high unit cost. Implementation costs are also currently high as most of the support systems, for example CAD/CAM and machine maintenance, are not yet as well developed as those for conventional machines. In the future, however, the reduced requirement for alignment of precision slideways coupled with the reconfigurability and standardisation of PKM devices should lead to reduced costs when compared with conventional machines of similar capacity.

7.4.4 *Disadvantages of PKMs*

Complex working volume

A conventional machine will typically have a working volume (the volume within which the tool is able to move) that is a simple cuboid. This will often be defined as fixed limits in each axis. For example, 1000 × 600 × 700 mm indicates a travel limit of 1000 mm in X, 60 mm in Y and 700 mm in Z. If additional rotational axes are present on the machine tool they will also have fixed limits. For example ±30° in A (rotation about the X-axis) and ±90° in C (rotation about the Z-axis). This simple way of defining the machine's working volume means that evaluating the machine's capability and determining whether a given component can be machined on it is fairly straightforward.

A PKM machine tool does not have actuators aligned to the conventional orthogonal (perpendicular) axes so the relationship between the limits of these actuators and the work volume is not direct as with a conventional machine. Also additional limits, such as pivot angle capability (maximum angle allowed at the actuators' rotational joints) and collision between the machine elements further increases the work volume complexity. Tools are currently being developed to address this issue of evaluating and utilising these complex workspaces.

Workspace issues

The term 'workspace' in this context refers to any characteristic that varies across the machine's working volume. Examples of these include accuracy, rigidity, velocity, acceleration and orientation (tilt) workspaces. Although these capabilities will also vary across the working volume of a conventional machine, with PKMs the variation is more complex as the actual structure of the machine changes considerably for each position.

The complexity of the workspace of these types of machine is an important issue to consider since it is no longer easy to assess the machine's capabilities for comparison with other PKMs or with conventional machines. In addition to this, the assessment of whether a given component can be machined and where it should be placed within the working volume of the machine also becomes complex.

Low rotational velocity

This is an issue that is frequently ignored when considering the advantages and limitations of PKMs. A pure parallel structure is designed to control six degrees of freedom of the tool tip and it is therefore a six-axis machine tool. In fact, only five of these axes are actually useful for machining as one axis is occupied by the tool axis and therefore redundant when a spindle is used, apart from changing the machine structure, but this topic is too complex to discuss here. This means that a PKM can manufacture complex components that require five-axis machining and reduce the number of set-ups for parts with multiple entry directions. However, the top platform and actuators have to move a great deal to achieve a relatively small degree of tilt. This is an inherent problem with parallel structures; to improve the rotational performance it is possible to reduce the distance between the top pivots and the tool tip. This would, however, significantly reduce the rotational accuracy. Some machine designers believe that the hybrid designs of PKMs with additional conventional axes could be a solution but this would also reduce some of the benefits of using parallel structures, because accurate alignment of the additional axes would be required.

Complexity and current lack of support systems

A PKM is a very complex machine and requires intensive control algorithms to achieve even simple motion. If a fault occurs on the machine it is often very difficult to identify the source of the problem. Additional operational issues include access limitations for maintenance, visibility restrictions and the difficulties of the integration of ancillary equipment such as tool changers and pallet shuttles. An operator may also find it hard to associate motion of the tool with changes in the machine structure. With the increased availability of powerful computing facilities at a low price and the research efforts being made into fault diagnosis and other support systems it should be possible to address and solve these complex issues in the foreseeable future, and PKMs may become the machines of the future. Further information concerning PKMs can be found on the 'Robotool' Web site at the Institute of Production Engineering and Machine Tools (IFW) at the University of Hanover (Web address: http://www.ifw.unihanover.de/robotool/). This includes an overview of the many existing machines along with links to related Web sites. The Variax Hexacenter has been described in some detail as it represents an entirely new approach to machine tool design. It is, as yet, too early to decide whether it heralds a new generation of machine tool designs. What is important is that an attempt has been made to look at old problems in new ways. Engineers must always keep an open mind if progress is to be made. However, change for the sake of change is equally counterproductive and a sensible balance must be struck.

The authors would like to take this opportunity to thank the following for their assistance in compiling this section and in granting permission for the use of their copyright material: Messrs Giddings and Lewis of America and Professor Nabil Gindy and the University of Nottingham.

7.5 Intelligent machines

The machine tool just discussed was based upon PKMs and as such represents what may lie ahead after further research and development. However, there are many 'intelligent' machines already on the market and in regular use. Traditionally human beings have relied upon the senses of touch, sight and sound to judge whether a machine is operating correctly. However, it is now possible to replace the human senses with piezo-electric sensors that can 'listen' to the process, probes, photo-electric devices and accelerometers built into the machine in order to detect when the operating parameters deviate from the ideal. Such information is fed back to the control computer and corrective adjustments are made. If this is not possible the computer may call up replacement tooling and replace the tools so that the machine can carry on working without human intervention. Should the problem be too serious for such remedial action the computer closes the machine down before serious damage can occur or expensive scrap components are manufactured. Provided the machine is linked with an automated load/unload system, it can work unattended under 'lights-out' conditions through the night. The machine has become *intelligent*.

Adaptive control, as such machine intelligence is called, is not new but is being constantly and increasingly developed as computers and their peripherals become more powerful. The aim of adaptive control is to recognise variations from the original operating parameters that may occur during a machining process, and make a suitable compensating response. Unless such a response is made the effect of the variations may lead to the production of components that are outside their limits of size (scrap) or may result in damage to the machine, the tooling or the workpiece. Adaptive control is essentially data feedback from various sensors built into or added to the machine. For example, a cutting tool becoming blunt can be identified by an increase in the power required to turn the machine spindle, this can be identified by torque monitoring. The temperature at the tool/workpiece interface will rise and this can also be monitored. The cutting forces imposed on the tool will also increase and these can be monitored. This information will result in the control computer taking the proper corrective action, usually by switching over to back-up tooling and alerting the operator accordingly so that the original tooling can be replaced. If the change to the back-up tooling does not solve the problem, the computer will close down the process until the cause of the problem can be investigated by the operator and corrected.

Surface sensing probes are also used for a number of purposes:

- To detect the presence of workpieces and to close down the process if no workpiece is found.
- To check the stock material sizes and automatically offset the work datums so that the finished part will lie within the boundaries of the stock material.
- To find the position of the datum surface so that the machine will automatically set itself.
- To speed up the process cycle by adjusting the feedrate to the prevailing conditions. For instance when no metal is found, as when the cutter is passing over a gap, the traverse rate can be increased and then reduced when metal is again detected. Furthermore, if the depth of cut is found to be greater than the amount expected (and programmed for) the traverse rate is automatically reduced.
- To perform in-process gauging so as to eliminate expensive inspection and sampling processes.

These are just a few of the applications of adaptive control and, as computing power continues to increase, more applications will inevitably become available. For example let us consider *neural networks*. This is the technology that is used in machines that can 'learn' by themselves. Neural networks are based upon closed loop, iterative processes. The machine learns like we do. It tries out a command and examines the outcome. If this shows an improvement it stores it in its memory. If the outcome is inferior it rejects it. It then tries to improve on a beneficial outcome and the process is repeated until the optimum performance is reached. Remember, when discussing turret punching, it was stated that the machine teaches itself to find the ideal stroke length for the punching operation to optimise the process. This is machine intelligence. Have you noticed that if you are numbering a sequences of notes on your computer, the computer takes over after the first one or two entries, anticipates your requirements and numbers and spaces the items for you. The computer has recognised what you are doing. It has learned. It has shown a level of intelligence. These are only simple examples but *neural networks* are becoming increasingly powerful, and computers and machines will become very much more 'intelligent' in a very short period of time.

7.6 Miniaturisation

Size reduction has a number of advantages:

- Less material is required resulting in less environmental pollution in its extraction and manufacture.
- The mass of products becomes less resulting in greater ease of acceleration and deceleration, also they become more portable and easier to handle.
- The volume of components is reduced so that they require less storage space.

It is only necessary to compare early computers that required a whole room or even a building to themselves and consumed prodigious amounts of electrical energy to drive them, with later, battery operated laptop (note-book) computer. Yet the laptop computer will have many times the computing power of its massive predecessor. Consider, also, a mobile phone of the latest generation. Mobile phones developed out of all recognition after a few years, and achieved the ability to receive text messages, colour pictures and graphics, and send and receive e-mails. They revolutionised instant and universal communication.

Yet such developments have only become possible by the adoption of microminiaturised components together with the technology to manufacture and assemble them. The chips for computer memory devices and central processor units start life as wafers cut from high-purity monocrystalline silicon. This is then 'doped' with trivalent or pentavalent elements to provide either p-type or n-type electrical characteristics.

7.6.1 *The manufacture of transistors and integrated circuits*

Devices with a *planar* configuration are manufactured so that the preparation of the various p-type and n-type layers and the metallised contacts are on one flat surface of the chip and are not on its sides or ends. The substrate is made from a relatively thick wafer of n-type silicon that has been doped to give it a low resistivity. The layers that are built up on this substrate are referred to as *epitaxial* layers. These are layers that grow onto the substrate surface as a continuation of the underlying crystal. The growth comes from the gaseous mixtures in which the wafer is heated to give p-type or n-type conducting layers or silicon dioxide insulating layers.

The fabrication of planar devices requires wafers with one surface lapped and polished to a high degree of flatness and surface finish. The size of the chip depends upon the number of components and circuits that are built upon it. A chip for a single transistor is only some 4 mm or 5 mm square and very many such devices can be made at the same time on wafers whose diameters lie between 100 mm and 150 mm. High-precision photographic processes produce the masks used during the fabrication of the device. Only one such device is drawn out and the camera takes a succession of photographs of the device on the same negative, moving by an increment equal to one chip spacing between each exposure. Thus the negative is covered in a pattern of chip masks suitable for printing photographically onto the wafer, which is coated with a photosensitive emulsion. The resolution required to reproduce the fine detail and intricacy of modern solid state devices precludes the use of white light, and ultraviolet light and laser light sources are used to make the exposures.

The memory chips and central processor chips of modern computers, especially laptop computers, have a very large number of individual components and circuits packed into a very small space. This defies 'mechanical' assembly and such devices are built up chemically. In some instances the insulating barriers between adjacent components are only a few molecules thick. The processes used in the manufacture of such devices are known as *nano-technology*. The latest devices are already reaching the limits of technology and it has been forecast that further development will not be possible without new technologies being developed. Already scientists are looking at *organic* technologies for computer memories.

However, computer science and application technology will not stand still. Even if the power of computers is reaching saturation point, software development will continue apace as new problems present themselves and new solutions are found. Just as the transistor and integrated circuit produced an ICT revolution only a short while ago, who is to say what the next development will be and what affect it will have on the human race and its work and leisure patterns.

7.6.2 *Micro-electromechanical system technology*

The micro-electromechanical (MEM) system is a rapidly developing technology that involves a combination of different engineering and scientific disciplines. It combines electronic engineering, mechanical engineering, materials science, optical technologies, physics, chemistry and bio-technology together in various combinations to solve wide-ranging problems involving devices varying in size from a few micrometres to a few millimetres. It uses micro-sensors, micro-actuators, micro-drive units and micro-controllers. Micro-electromechanical systems are already widely used in industry, aerospace, bio-medicine, surgery and agriculture. In fact, any situation where size and weight is important and micro-miniaturisation is the only solution. For example surgical implants such as heart pacemakers, and artificial limbs with miniaturised computers and micro-actuators that can be linked to and controlled by the user's natural neural system, are already in use.

7.7 Security in e-manufacturing and e-commerce

It has been stated earlier in this book that one of the limiting factors in the exchange of information over the Internet is maintaining security and preventing loss of intellectual copyright. This is by no means a new problem and codes and ciphers have been used for thousands of years to avoid the interception of communications. The famous enigma machine of the Second World War was originally invented for commercial use before being used by the German military for the coding of commands.

The Internet is used increasingly by all and sundry ranging from private individuals to 'big-business' to carry out an equally wide range of transactions. Regrettably the use of computer transactions linked through the public telecommunications network worldwide lays itself open to fraud. Currently, it is estimated that more than $10 billion worth of data is stolen annually in the USA alone. Therefore Internet security is a major problem that has to be solved.

Existing and proven security technologies are based upon authentication provided by passwords/PINs, encrypytion and digital signatures.

Passwords and PINs are regularly used for many transactions by individuals and companies, however, they do have a number of disadvantages, as is proven by the current level of computer fraud. Passwords and PINs should be memorised but, because so many are now required, they are either written down or are chosen to be easy to remember. Therefore they are open to theft by a computer 'hacker'. Often passwords can be guessed by associating places, names, nicknames and dates with the owner. Electronic eavesdropping is also becoming increasingly sophisticated so that passwords and PINs are no longer as secure as they originally were.

Encryption is widely used by companies rather than by individuals and the process involves sophisticated mathematical conversion techniques. Encryption is an important technology in e-commerce. There are basically two systems:

- *Symmetrical cryptography* where the same, secret 'key' is used to encrypt and decrypt. This has the disadvantage that for each user to have unique access to an item of information the sender and the recipient must each have a copy of the same key. Where a very large number of clients are involved an equally large number of individual 'keys' are required. Each 'key' must be privy to the sender and recipient so that no one else can access the information.
- *Assymetric cryptography* uses a pair of linked 'keys', one public and one private. Only the recipient has a private key and the public 'key' is common to all senders who wish to communicate with the private key holder. This substantially reduces the number of different 'key' patterns required.
- *Digital signatures* are an electronic means of verifying the authenticity of a computer transaction. Digital signature systems must have two functions. First, they must authenticate the signatory so that authorship cannot be denied. Secondly, the document that has been digitally authenticated must be incapable of alteration after signature.

Unfortunately there are always criminals who have the technology and skills to break into apparently secure systems so that there is a continuous race between the developers of new codes and systems and the code-breakers who penetrate security systems. Current technology prevents a third party from intercepting an online transmission, however, it is not proof against imposters accessing information, if the imposter uses a stolen PIN. The system has no means of identifying who used the PIN to issue the request for information or, in the case of a bank account, a sum of money.

One emerging technology that is attempting to overcome the breaking of security systems is *biometrics*. This is based on the ability of electronic equipment to read fingerprints or retinal prints that are unique to each human being. Inevitably, within a few years, biometrics will replace or supplement existing methods of authentication for sensitive transactions. Such improvements in online security will inevitably lead to an even greater and more rapid expansion of e-commerce and e-manufacture, for example agile manufacturing employing Web-based technologies.

The importance of the role of manufacturing in global markets cannot be overestimated. For manufacturing to succeed it must acquire new levels of flexibility and responsiveness: *agile engineering*. Already the World Wide Web (WWW) is the most widely used and visible component of the Internet because of its multimedia capability. This has resulted in its adoption for Web-based agile manufacture. Already research in China is addressing the challenges of improving cooperative design and manufacturing responsiveness by establishing Web-based systems to support application integration and enterprise-wide communication.

7.8 Parts-sourcing in a global market

There is an old saying that there is no future in reinventing the wheel. Therefore every designer uses standard components wherever possible. This ensures availability at minimum cost and guaranteed quality not only in the initial product build but also for future replacements during maintenance and repair. To do this on a national basis is not too difficult but, on a global basis, it can become a time-consuming and, therefore, costly exercise. Thus *parts-sourcing in a global market* is an important development of computer technology. It involves multimedia 3D modelling and the development and use of 3D Web sites. It provides a global market both for a firm's products and from which essential products can be readily sourced. The collaborative re-use of design and manufacturing data is one way in which e-commerce can significantly reduce costs and lead time for new products in the demanding environment of global markets. Earlier in this book we looked at the technique of 3D modelling and the time is not far distant when all manufacturers of standard components and sub-assemblies will post 3D computer-generated models of their products on the Internet. These models will be analysed by a 3D Internet-based search engine and relevant feature indices will be stored in a database. Designers will be able to upload a computer-generated 3D model of a required part, and the search engine will attempt to achieve a match or a near match with similar products in its database. The designer is then presented with a range of similar parts and he or she will be able to modify the design to accept the closest standard available product. The designer can then order the required part over the Internet. Alternatively the designer can contact the supplier of the nearest available similar part for a quotation to supply a variant of its existing product to exactly match the requirements of the new design.

7.9 Green process planning technology

Through the efforts of the media, people have an enhanced perception of environmental (green) issues and, as far as possible, demand a pollution-free environment in which to work and in which to take their leisure. Most companies have come to realise that environmental issues are not only an important public relation exercise but can result in costs savings particularly in the use of energy saved. The key issues to be considered at the design stage can be summarised as material selection, processing, consumer use and eventual disposal.

7.9.1 *Materials*

Wherever possible the materials used should come from renewable and sustainable sources or they should be the outcome of recycling. Even then such materials should be used in the minimum-possible quantities. This not only reduces the volume of waste disposal but also reduces the amount of energy used to process and transport the material. Where traditional materials, such as metals, have to be used, every effort must be made to minimise the pollution caused in their extraction. Unfortunately fossil fuels are often an important part of the chemical reactions necessary to extract metals from their mineral ores. Therefore, again, such materials should be used in the least-possible quantities and sourced from plants that have been modernised with environmental issues in mind. It is often suggested that battery electric cars would reduce pollution. This is true on a local basis but it must be remembered that the extraction of lead for their batteries and the eventual disposal of this material would itself cause massive pollution problems. Further, additional electricity would have to be generated to charge the batteries. Again this would lead to further pollution unless renewable energy sources such as hydro-electricity, tidal-generation or wind powered generators can be developed on a much bigger scale.

7.9.2 *Processing*

Processing should also be designed with environmental issues in mind. For example, minimum energy requirement, recycling of waste, removal of particulate pollutants from the atmosphere and the reduction of noise pollution, because noise is wasted energy. The Health and Safety at Work Act expects employers to provide a safe and healthy environment in which their employees can work. It also expects employers to take a responsible attitude to keeping the environment of the community in which their plant is sited free from all forms of pollution that might be generated from their manufacturing activities. This particularly applies to the discharge of toxic and noxious chemicals and particulates into the atmosphere and into local watercourses.

7.9.3 *Consumer use*

It is not sufficient to design a product that is sourced from renewable and sustainable material sources and to employ environmentally-safe manufacturing processes, if the final product is not environmentally friendly in its day-to-day use. Most domestic appliances aim to use minimum energy and operate quietly. Washing machines are designed to use minimum amounts of water and discharge minimum amounts of cleansing agents down the drain. Washing powders themselves are compounded to be as environmentally friendly as possible and work effectively at minimum temperatures. Cars use catalytic converters to minimise the discharge of noxious exhaust gases that cause global warming. At the same time, engine designers have produced engines that have 'lean burn' characteristics so as to keep exhaust discharge to a minimum through a reduction in the amount of fuel used. To a large extent this has been made possible by the use of dedicated engine management computers that maintain maximum performance with minimum fuel consumption.

7.9.4 *Disposal*

Waste disposal is an ever-increasing problem in the developed world. To some extent the solution of one problem brings another in its wake. For example, in the days of open coal fires much household waste could be burnt. With the banning of such fires in many areas all household waste has had to go in the bin for disposal by the local authority. At the same time we have become more health conscious and expect our food to be hygienically wrapped. The development of supermarkets has resulted in most products being pre-packed so that they can be taken from the shelf in known quantities at a standard price for the benefit of the 'till operator'. There is no longer time to weigh and wrap individual items. Glass and plastic containers can be recycled but most waste goes into 'landfill' sites that have their own problems if not properly managed.

7.9.5 *Green process planning technology*

Having made the case for green process planning, it is now necessary to investigate what can be done about it. Essentially the problem is multidisciplinary involving most branches of science and technology. Process planning links the design process, the deployment of manufacturing resources and quality control. Green process planning has similar goals but also targets energy saving, the use of renewable materials and the reduction of all forms of pollution to achieve *sustainable manufacture*. This depends upon a great deal of multidisciplinary collaboration and information sharing that can only be achieved by setting up specialist Web sites on the Internet together with specialist search engines to harvest, sieve and analyse the available information. Web-based collaboration will depend upon the sharing of product models and the sharing of definitions and objectives within a framework of concurrent engineering. Collaborative design should be achieved by esoteric philosophies such as serialisation, modularisation and customisation of cell sub-products by computer networking and the use of artificial intelligence.

It should be clear by now that computer hardware may be reaching the limits of development – except in size – within known electronic technology. There is a never-ending range of opportunities for the development of software and applications in e-manufacture and e-commerce if a sustainable, global manufacturing community is to be achieved that seeks to satisfy the growing demands and expectations of the consumer.

ASSIGNMENTS

1. Discuss the economic, sociological and technological impact of developments in e-manufacturing on the future of global manufacturing.

2. Discuss the economic and technological criteria that need to be considered when assessing a new process and describe how you justify its adoption to the board of your company.

3. Compare and contrast the advantages and limitations of parallel kinematic mechanisms with conventional machine tool construction.

4. Discuss the meaning of the term 'machine intelligence', citing and describing an example with which you are familiar.

5. Discuss the advantages and limitations of miniaturisation as applied to manufactured components and cite examples where such miniaturised components can be used to advantage.

6. Discuss and describe methods for making e-manufacture and e-commerce secure so as to limit the damage caused by industrial espionage and the reluctance on the part of the business community to use the Internet for sensitive information.

7. Discuss the need for part-sourcing in a global market, the advantages and limitations of such a system, and the design criteria for a suitable 3D Internet-based search engine.

8. Discuss the importance for companies to adopt green process planning technology in the current and future industrial and political/social environments. Describe the implementation of such a strategy at any company with which you are familiar. Discuss any economic and technological advantages and limitations that have accrued from the adoption of such a strategy.

Appendix

This Appendix is included for the benefit of those students who wish to apply the principles and techniques expounded in the text of *E-manufacturing* to mechanical engineering and engineering materials.

Manufacturing Technology, Volume 1, 3rd edition, R. L. Timings

Contents

Chapter 3 Toolholding and workholding

3.1 Principles of location and restraint
3.2 Practical locations
3.3 Practical clamping
3.4 Drilling machine: toolholding and workholding
3.5 Milling machine: toolholding and workholding
3.6 Lathe: toolholding and workholding
3.7 Grinding machine: toolholding and workholding

Chapter 4 Kinematics of manufacturing equipment

4.1 Generating and forming of surfaces
4.2 Requirements of a machine tool
4.3 The basic structure
4.4 Power transmission
4.5 Guidance systems
4.6 Positioning systems
4.7 Machine-tool alignments
4.8 Machine-tool accuracy
4.9 Setting accuracy

Chapter 5 Numerical control part programming

5.1 Background
5.2 Applications
5.3 Advantages and limitations
5.4 Axis nomenclature
5.5 Control systems
5.6 Data input
5.7 Program terminology
5.8 Program formats
5.9 Canned cycles
5.10 Tool length offset
5.11 Cutter compensation
5.12 Tool retrieval
5.13 Workholding
5.14 Part programming

Chapter 6 Assembly

6.1 Design of components
6.2 Batch size
6.3 Accuracy of fit
6.4 Toleranced dimensions
6.5 Standard systems of limits and fits
6.6 Selection of tolerance grades
6.7 Interchangeability
6.8 Materials
6.9 Joining methods

****

Manufacturing Technology, Volume 2, 2nd edition, R. L. Timings and S. Wilkinson

Contents

Chapter 5 Plastic moulding processes

5.1   Plastic moulding materials
5.2   Polymerisation
5.3   Forms of supply
5.4   Positive-die moulding (thermosetting plastics)
5.5   Flash moulding
5.6   Transfer moulding
5.7   Compression-moulding conditions
5.8   Injection moulding
5.9   Use of inserts
5.10   Extrusion
5.11   Selection of moulding process
5.12   Justification

Chapter 6 Cutting processes

6.1   Sheet metal
6.2   Process selection (introduction)
6.3   Process selection (surface geometry and process)
6.4   Process selection (accuracy and surface finish)
6.5   Process selection (tooling and cutting costs)
6.6   Process selection (cutting tool materials)
6.7   Capstan and turret lathes
6.8   Automatic lathes
6.9   CNC turning centres
6.10   Screw-thread production (lathes)
6.11   Screw-thread production (milling)
6.12   Screw-thread production (grinding)
6.13   Production tapping
6.14   Gear-tooth production (forming)
6.15   Gear-tooth production (generation)
6.16   Broaching
6.17   Centreless grinding
6.18   Transfer machining
6.19   Electrical discharge machining – principles
6.20   Electrochemical machining
6.21   Chemical machining
6.22   Laser cutting
6.23   Justification

Chapter 7 Measurement and inspection

7.1   Kinematics of measuring equipment
7.2   Instrument mechanisms
7.3   Magnification
7.4   Angular measurement
7.5   Laser measurement
7.6   Probing

****

Engineering Materials, Volume 1, 2nd edition, R. L. Timings

Contents

****

Engineering Materials, Volume 2, 2nd edition, R. L. Timings

Contents

Part A Metallic materials

Index